Singapore

Singapore, renowned for its prosperous modern industrial economy, also enjoys a more controversial reputation through its strong political rule and pragmatic policies. Yet this strong hegemonic state achieves effective rule not just from repressive policies but also through a combination of efficient government, good standard of living, tough official measures and popular compliance. Souchou Yao looks at the reasons behind such paradoxes, examining key events such as the caning of American teenager Michael Fay, the judicial ruling on fellatio and unnatural sex, and Singapore's 'war on terror' to show the ways in which the State manages these events to ensure the continuance of its power and ideological ethos. Subject areas discussed include:

- leftist radicalism and communist insurgency
- nation-building as trauma
- Western 'yellow culture' and Asian Values
- judicial caning and the meaning of pain
- the law and oral sex
- food and the art of lying
- cinema as catharsis
- Singapore after September 11

Singapore traverses several fields of study, taking up ideas and frameworks from philosophy, psychology, political science, cultural studies and anthropology in order to tell the larger 'truth' about the Singapore state. As such, this lively and well-written book will appeal to anyone interested in Singapore and Asian studies alike.

Souchou Yao is Senior Lecturer in Anthropology at the University of Sydney. He has published widely in international journals and is the author of *Confucian Capitalism: discourse, practice and the myth of Chinese enterprise*.

Asia's Transformations
Edited by Mark Selden
Binghamton and Cornell Universities, USA
The books in this series explore the political, social, economic and cultural consequences of Asia's transformations in the twentieth and twenty-first centuries. The series emphasizes the tumultuous interplay of local, national, regional and global forces as Asia bids to become the hub of the world economy. While focusing on the contemporary, it also looks back to analyse the antecedents of Asia's contested rise. This series comprises several strands:

Asia's Transformations aims to address the needs of students and teachers, and the titles will be published in hardback and paperback. Titles include:

Debating Human Rights
Critical essays from the United States and Asia
Edited by Peter Van Ness

Hong Kong's History
State and society under colonial rule
Edited by Tak-Wing Ngo

Japan's Comfort Women
Sexual slavery and prostitution during World War II and the US occupation
Yuki Tanaka

Opium, Empire and the Global Political Economy
Carl A. Trocki

Chinese Society
Change, conflict and resistance
Edited by Elizabeth J. Perry and Mark Selden

Mao's Children in the New China
Voices from the Red Guard generation
Yarong Jiang and David Ashley

Remaking the Chinese State
Strategies, society and security
Edited by Chien-min Chao and Bruce J. Dickson

Korean Society
Civil society, democracy and the state
Edited by Charles K. Armstrong

The Making of Modern Korea
Adrian Buzo

The Resurgence of East Asia
500, 150 and 50 Year perspectives
Edited by Giovanni Arrighi, Takeshi Hamashita and Mark Selden

Chinese Society, 2nd edition
Change, conflict and resistance
*Edited by Elizabeth J. Perry and
Mark Selden*

Ethnicity in Asia
Edited by Colin Mackerras

The Battle for Asia
From decolonization to
globalization
Mark T. Berger

**State and Society in 21st Century
China**
*Edited by Peter Hays Gries and
Stanley Rosen*

Japan's Quiet Transformation
Social change and civil society in
the 21st century
Jeff Kingston

Confronting the Bush Doctrine
Critical views from the
Asia-Pacific
*Edited by Mel Gurtov and
Peter Van Ness*

**China in War and Revolution,
1895–1949**
Peter Zarrow

**The Future of US–Korean
Relations**
The imbalance of power
Edited by John Feffer

Working in China
Ethnographies of labor and
workplace transformations
Edited by Ching Kwan Lee

Singapore
The State and the culture of
excess
Souchou Yao

Asia's Great Cities. Each volume aims to capture the heartbeat of
the contemporary city from multiple perspectives emblematic of the
authors' own deep familiarity with the distinctive faces of the city, its
history, society, culture, politics and economics, and its evolving position in national, regional and global frameworks. While most volumes
emphasize urban developments since the Second World War, some pay
close attention to the legacy of the *longue durée* in shaping the contemporary. Thematic and comparative volumes address such themes as
urbanization, economic and financial linkages, architecture and space,
wealth and power, gendered relationships, planning and anarchy, and
ethnographies in national and regional perspective. Titles include:

Bangkok
Place, practice and representation
Marc Askew

Beijing in the Modern World
*David Strand and Madeline Yue
Dong*

Shanghai
Global city
Jeff Wasserstrom

Representing Calcutta
Modernity, nationalism and the
colonial uncanny
Swati Chattopadhyay

Hong Kong
Global city
Stephen Chiu and Tai-Lok Lui

Singapore
Wealth, power and the culture of
control
Carl A. Trocki

Asia.com is a series which focuses on the ways in which new information and communication technologies are influencing politics, society and culture in Asia. Titles include:

Japanese Cybercultures
*Edited by Mark McLelland and
Nanette Gottlieb*

**The Internet in Indonesia's New
Democracy**
David T. Hill and Krishna Sen

Asia.com
Asia encounters the Internet
*Edited by K. C. Ho, Randolph
Kluver and Kenneth C. C. Yang*

Chinese Cyberspaces
Technological changes and
political effects
*Edited by Jens Damm and
Simona Thomas*

Literature and Society is a series that seeks to demonstrate the ways in which Asian Literature is influenced by the politics, society and culture in which it is produced. Titles include:

**The Body in Postwar Japanese
Fiction**
Edited by Douglas N. Slaymaker

**Chinese Women Writers and the
Feminist Imagination,
1905–1948**
Haiping Yan

Routledge Studies in Asia's Transformations is a forum for innovative new research intended for a high-level specialist readership, and the titles will be available in hardback only. Titles include:

1. **The American Occupation
   of Japan and Okinawa***
   Literature and memory
   Michael Molasky

2. **Koreans in Japan***
   Critical voices from the
   margin
   Edited by Sonia Ryang

Critical Asian Scholarship is a series intended to showcase the most important individual contributions to scholarship in Asian Studies. Each of the volumes presents a leading Asian scholar addressing themes that are central to his or her most significant and lasting contribution to Asian studies. The series is committed to the rich variety of research and writing on Asia, and is not restricted to any particular discipline, theoretical approach or geographical expertise.

Southeast Asia
A testament
George McT. Kahin

**Women and the Family in
Chinese History**
Patricia Buckley Ebrey

China Unbound
Evolving perspectives on the
Chinese past
Paul A. Cohen

China's Past, China's Future
Energy, food, environment
Vaclav Smil

**The Chinese State in Ming
Society**
Timothy Brook

Singapore

The State and the culture of excess

Souchou Yao

Routledge
Taylor & Francis Group

LONDON AND NEW YORK

First published 2007
by Routledge
2 Park Square, Milton Park, Abingdon, Oxon, OX14 4RN

Simultaneously published in the US
by Routledge
270 Madison Avenue, New York, NY 10016

Routledge is an imprint of the Taylor & Francis Group, an informa business

Typeset in Baskerville by Swales & Willis Ltd, Exeter
Printed and bound in Great Britain by
Antony Rowe Ltd, Chippenham, Wiltshire

British Library Cataloguing in Publication Data
A catalogue record for this book is available
from the British Library

Library of Congress Cataloging-in-Publication Data
Yao, Souchou.
Singapore : the state and the culture of excess / by Souchou Yao.
p. cm.
Includes bibliographical references and index.
1. Singapore – Social conditions – 21st century. 2. Singapore – Social
policy. 3. Singapore – Politics and government – 1990-. I. Title.
HN700.67.A8Y36 2007
306.2095957 – dc22
2006022762

ISBN10: 0-415-41711-2 (hbk)
ISBN10: 0-415-41712-0 (pbk)
ISBN10: 0-203-96551-5 (ebk)

ISBN13: 978-0-415-41711-2 (hbk)
ISBN13: 978-0-415-41712-9 (pbk)
ISBN13: 978-0-203-96551-1 (ebk)

Contents

Preface and acknowledgements

Like baby boomers with the Beatles, my anthropologist's career is wonderfully caught up with the writings of Marshall Sahlins. His *Stone Age Economics* guided my shifting of camp from economics to anthropology, and reading *Culture and Practical Reason* drove home to me for the first time that culture and material logic always sit cheek by jowl in a single realm of social processes. As a postgraduate student in Adelaide I even shared a taxi with the great man on the way to a departmental lunch a couple of decades ago. So I followed him from *Stone Age Economics*, the sharp insights into state formation in Indonesia and the culture and religion of Bali, to his acerbic rebuttal of postcolonial anthropology's excesses in *How 'Natives' Think*, and finally to his *Apologies to Thucydides: Understanding History as Culture and Vice Versa*. *Apologies* is philosophic rumination in the best sense of the phrase. Thucydides' account of the Peloponnesian War opens up, in a fertile intellect, the political intricacies of the Polynesia War and the contest for power in the Fijian Islands, the baseball triumph of Bobby Thomson for the 1951 Giants and the Elàn Gonzalez custody struggle. The richness of subjects and reflections shows up the primary interconnectedness of individual and national-tribal histories, political power and personal wishes, in the making of the world. *Apologies*' philosophic turn shows how organically engaging anthropology can be written. I am somewhat shamefaced about this name-dropping to justify my lame imitation of Sahlins's great works. But it is hard not to take up the lesson: that life's events are both particular and universal and that, try as it may, a society is never bound by its parochial passion. If this is true of pre-colonial Polynesia, it is also true of the small island republic of Singapore in the tropical South China Sea.

Over the last two decades, parochial passion has all but taken over the state discourses and much of the local debates about the position xii Preface and acknowledgements

of Singapore in the transnational world. This has taken place in the midst of expanding economies and an aggressive courting of Western capital. 'Confucian capitalism', Asian and Confucian Values, and Singapore anti-West posturing all seemingly articulate the country's new needs and agendas. Singapore is the sign of New Asia: confident, prosperous, and prepared to challenge the cultural hegemony of the West. And inevitably the position would have Singapore revising some of the ideas on individual freedom and liberal democracy inherited from Europe. The Singapore State's attempt to rewrite its new status, and present its case to the people and the international world, clearly calls for a reasoned and reasonable rebuttal.

Singapore is a place of many paradoxes: a society of First World living standards, yet it is ruled by harsh state measures and pragmatic policies reminding one more of the practices of a Third World nation; a society with an advanced economy, yet its liberal-democratic standards fall short of similar 'development'. Its political leaders have a talent for efficient administration, but they are seemingly inflicted by the 'tragic flaw' of self-possession. The book is an examination of this self-possession: how it settles the State in the assumptions of its ideas and actions, and blinds the State to critiques and alternative visions. Here I think of critiques in the Enlightenment sense, not as attack and denigration, but as a project that opens up the mind's deficiency and lack of self-knowledge and, in more modern terms, the mystifying effects of ideology. Writing from Australia, I sometimes find it hard to recall that cultural critique can have a different meaning, and consequences for the practitioner, in Singapore. Be that as it may, it is impossible to sustain a project like this if the writer does not to an extent empathize and identify with the subject he or she is writing about. For all my probing analysis, I cannot write about Singapore as if it were a land of ruthless repression and its people hostages to the State. This no doubt does not go far enough for Singapore's critics; neither will it satisfy the supporters of its ruling authorities. Singapore's social and economic achievement, not the least its charismatic National Father, Lee Kuan Yew, require us to be fair. But this is quite different from a mindless acceptance of everything the PAP State has said and done. Truth and fairness also demand that we look at the State's arrogation of power and the poverty of moral imagination that is also the Singapore Story.

This book began its life while I was a fellow at the Institute of Southeast Studies, Singapore, from 1993 to 1996. I kept a journal, collected newspaper cuttings and did a great deal of 'weekend field-work'; these and numerous later visits to Singapore provided the

materials of the book and informed my understanding of the events
it describes. Conferences and lectures in various institutions offered
not only a welcome break from the writerly desperation, but also
conversations that sharpened my thinking. I would like to mention in
particular the Department of Sociology, Kassel University, Germany,
and the Asian Research Institute (ARI), National University of Singapore, in this regard. Chapter 6 was first written for the 'Politics of
Food' conference held in July 2003 at the ARI and organized by the
Geography and Chinese departments of the National University of
Singapore.

I have made each chapter more or less complete in itself. Each deals
with a particular event and my analysis of it. Chapters 1 and 2 bring up
the history of the struggle for national independence, the violence and
political volatility of which produced what I called the 'totalitarian
ambitions' and the 'culture of excess' of the Singapore State. They
should be read first. Chapters 6 and 7 move away from the State and
document the effects of its oppressive measures on the everyday and
popular imagination. Chapter 8 traces Singapore's response to the
September 11 terrorist attack, and the significance of Lee Kuan Yew's
eightieth birthday, and provides a synthetic view, less a conclusion, of
Singapore in the early twenty-first century.

In the text, 'State' refers to the Singapore State, and 'state' the
political formation in general. I hope this avoids any confusion, since
the terms in both senses appear together in some chapters. The use
of the feminine personal pronoun as in the current editorial protocol
causes some problems. Singapore state leaders in the cabinet are all
men, and women prisoners are not sentenced to be caned; in these
cases addressing any of them as 'she' is confusing and dissonant. So
I have resorted to the use of 'he'; and to refer to the Singapore State
and nation as 'it' seems more consistent with the present-day gender
politics.

Given the long fermentation of this book, it is hard to acknowledge
all the people who have helped me along the way. Nonetheless I would
like to mention the contributions and encouragement of the following,
alphabetically: Michael Barr, Mark T. Berger, Ashley Carruthers,
Chen Kuan Hsin, Chua Beng Huat, John Clark, Tom Ernst, Develeena
Gosh, Lisa Low, Carl Trocki, C. J. W.-L. Wee, Anna Yeatman. Michael
Carter generously shared with me his insight into the works of Kojève;
Chapter 8 on the reading of Singapore at the 'end of history' is due
partly to his inspiration, though he is not responsible for any mistakes
it may contain. My wife, Simryn Gill, has her own busy schedule and
preoccupations as an artist, but clearly we have in imperceptible ways

drawn influences from each other's work. Outside the university circle, I have the benefits of the friendship of Yap Lim Sen, Beth Yaph, Ken Yeh and Carol Marra – fellow Malaysians all. Yap Lim Sen on reading the manuscript with his friends at the Ipoh Turf Club christened it *Singapore on the Couch*. I wish I could keep the title, for the imagery of a nation-state in mental turmoil seeking cure is arguably the heart of the book's engagement. Finally I thank Stephanie Rogers for efficiently seeing the manuscript through to publication, and Mark Selden for his support and the legendary promptness of his responses to emails.

All money values are in Singapore dollars, unless otherwise stated. In June 2006 the exchange rate was SG$1.58 to US$1.

1 The magic of the Singapore State

[For the moral] act to be everything it should be, for the rule to be obeyed as it ought to be, it is necessary for us to yield, not in order to avoid disagreeable results or some moral or material punishment, or to obtain a certain reward; but very simply because we must, regardless of the consequences our conduct may have for us. One must obey a moral precept out of respect for it and for this reason alone.

Emile Durkheim, 'The Science of Morality'

Nation and the Sick Father

In 1965 Lee Kuan Yew – Singapore's Founding Father and first Prime Minister – had what one writer describes as a 'minor breakdown'.[1] It happened after Singapore's expulsion from the Federation of Malaysia, a political merger that Lee had hoped would build a common economic union and realize his vision of a 'Malaysian Malaysia' based on meritocracy rather than ethnic – Malay – preferences.[2] The separation was for Lee personally and politically traumatic. He had broken down in tears at the press conference where he announced the news. The collapse of the merger left him 'drained physically, emotionally and mentally' and he had to seek six weeks' retreat in government barracks in Changi to recover.[3] But personal disappointment was not the only reason for his breakdown. In the preceding months, he had worked feverishly in negotiating with the sensitive Malay leadership in Kuala Lumpur to secure a viable future for the Chinese-dominated city-state. For a man given to emotional restraint and tough-mindedness, the separation of Singapore from the Federation nevertheless 'opened the flood gates' that had been walled up for years.[4] He was, in his own words, 'emotionally overstretched' and 'close to physical exhaustion'; and the separation 'weighed [him] down with a heavy sense of guilt' for having failed his supporters and allies.[5] To help him sleep his doctor had prescribed sedatives, and pep pills to keep awake to face the day. Taking these drugs in a condition of nervous exhaustion had a debilitating effect and left Lee in a dark and volatile mood:

> Some in Lee's circle ... felt that commonsense advice had been neglected because of a pharmacological bias to his doctor's training. The drugs had an innocuous enough effect when Lee could see his way through situations, but under the enormous strain of recent events their impact was curious and unpredictable. One moment Lee could be smiling, offering Tunku a brittle picture of acceptance, even some sort of pleasure. The next moment when he was near people with whom he could allow himself to relax – colleagues, selected foreign journalists, subordinates – he would burst into tears or pour forth a torrent of emotion-laden words, recollections, predictions.[6]

To Lee's outbursts of tears and emotions, one writer would add fear of assassination.[7]

For all that, it was not in Lee's character to succumb, either to his enemies or to what he himself might regard as 'personal weakness'. During his recuperation, and having given up his heavy work schedule, he continued to travel from Changi to attend weekly cabinet meetings in the city. On the night of 30 September he received news of a military coup in Indonesia led by General Suharto; it triggered another bout of sleepless brooding. As he remembers:

> I did not sleep well. Choo [Lee's wife] got my doctors to prescribe tranquillizers, but I found beer or wine with dinner better than the pills. I was then in my early forties, young and vigorous; however hard and hectic the day had been, I would take two hours off in the late afternoon to go on the practice tee to hit 50–100 balls and play nine holes with one or two friends. Still, I was short of sleep. Late one morning, when the newly arrived British high commissioner, John Robb, had an urgent message for me from his government, I received him at home lying in bed, physically exhausted.[8]

The British Prime Minister, Harold Wilson, was informed of his condition and expressed concerns. Lee wrote back to assure him:

> Do not worry about Singapore. My colleagues and I are sane, rational people even in our moments of anguish. We weigh all

possible consequences before we make any move on the political chessboard.... Our people have the will to fight and the stuff that makes for survival.[9]

The reply brought a certain normality to his conditions. For Lee had always worked at a high-pitched pace, impatient to get on with the task at hand, and intolerant of sloppiness and the weak-mindedness that he encountered in the administration. In his public speeches, he tended – as he still does – to 'lecture', to test his ideas before the audience and to cajole them with his compulsive insights. In the tumultuous 1960s and 1970s, both the Singapore nation and the People's Action Party (PAP) under his leadership were fighting for survival. It was scarcely the time for calm repose and reconciliatory gestures. Confronting the radical labour unions at home and critics abroad, Lee was in a combative mood. He was ready to strike at his enemies and to engage in the kind of brilliant rhetorical jousting for which he is renowned. Passionate and full of urgency, Lee's speeches in the period were a mixture of political resolve and moral anger. In the attempt to keep one step ahead of his enemies, and in pushing himself – and the nation he led – to the limit, a certain 'excessiveness' in his speeches and actions became the trademark of Lee's political style.

Thus in the early years of 'national struggle', it soon became clear where Lee's restless mind would take his listeners. In 1962 Lee spoke to the Malayan students in London:

If we lose, fritter away the next decade that we have and not make preparation for our take-off into the industrial age, then we may well have to regret it....

We have got to make sure that the capital we have accumulated is put to good use, that in ten years we take one stride forward, in twenty years we enter the industrial age and in thirty years definitely, we are an emerged nation, not an emerging one. Because, definitely in thirty years, we are going to have an emergent China.[10]

The lesson is clear: people must forever try harder, for all strivings are haunted by their potential failure, just as they risk being sabotaged by the complacent among the ranks and by foes out to dismantle the State. And the chilling vulnerability in Lee's view of things would have him imagine the emergence of powerful competitors who would cast a dark shadow over Singapore's future. In 1964, a year after the formation of Malaysia,[11] he expressed his fear:

> One day, God forbid, not too soon, in [Indonesia]...some order will be restored in place of chaos, and they will begin to move forward. Any time now, it is estimated that the Chinese government can explode a nuclear device. Any time now, the Indians are going to set up jet fighter factories.... The moment one of these countries outstrips Malaysia in the human material comforts of life... [Malaysia] must go asunder.[12]

In a speech to Malaysian students a month later, he gave the same warning:

> One day, I don't know when – 10, 15, 20 years – one of these three countries will overtake us in terms of material wealth and power; either Indonesia or China, or India. And if before then, we have not yet welded the three communities into a national identity, then I say we must come unscrambled....
> Time is not on our side, as far as this crucial issue is concerned.[13]

If all this sounded like something of an overwrought imagination, nonetheless Lee's near-apocalyptic vision carried, as always, a great deal of political realism. In the two decades after the end of the Second World War, the 'wind of change' sweeping across European colonies in Asia and Africa heralded the call for decolonization and nationalist struggles. Lee, once assured that the war-weary Labour Government in London was ready to grant independence, turned his energy to confronting the problems facing the young Malaysian Federation of which Singapore had become a part in 1963.[14] 'An Indonesia in chaos', a 'jet fighter-manufacturing India' and 'a China with a nuclear exploding device' warned of the prospects of regional conflict brought about by the arms race, ethnic violence and economic rivalry, a conflict into which the young Malaysian nation (and the Singapore State) would be inevitably drawn. It belongs to the geopolitics of the time to want to express one's concern with such deep pessimism; and in this sense Lee's excessive imagination is more than a product of his pathological condition. He was at the time facing a 'nation in crisis'; his passionate responses signalled a mode of action that went beyond what was normally expected 'under ordinary circumstances'. But neither Lee – his personality, intellect and political instincts – nor the circumstances facing Singapore were 'ordinary'. The demands of radical labour unions and Chinese students, the sensitivities of Malay leaders in Kuala Lumpur, Malay–Chinese racial tension at home, and economic uncertainty created by the eventual British military withdrawal all called for tough measures, just as they created a great deal of anxiety.

So Lee had fallen ill under these harrowing circumstances. But to the young nation his infirmity did not signal bodily weakness so much as offer unmistakable evidence of his wilful determination to overcome the odds stacked against him and the nation he led. Besides, a nation in crisis is an opportunity to demonstrate the iron resolve of 'men of destiny'. Lee, highly strung and too restless for sleep, calls up the figure of one fallen ill from worrying. In those difficult years, Lee's break-down spoke to the people with an ineluctable message. Not of feeble constitution or personal weakness, the National Father had fallen sick by taking the burden of the crisis-ridden world on his shoulders.

This moral attitude, this 'culture' of sacrifice that puts the national community above oneself, is indeed the Father's gift to his people. Over the following decades, in the massive social and economic transformation of Singapore, a crucial moralism – a probity of values and actions – was carefully guided to coalesce into national culture and the dominant ethos of the State.

Like the Father, the Singapore nation must cultivate a robust moral sense in its enterprises. As Singapore builds wealth from industrial capitalism, prosperity and the freedom of the market are also seen as posing a danger in producing self-possessing national subjects who care only for themselves and their private enjoyment. In the State's current view, economic richness also carries moral lassitude, which has infected the West, but is now coming dangerously to the Eastern shore. There is a distinct postcolonial gesture in this: since capitalist modernity tends to nurture self-willing individualism in the West, an Asian state like Singapore must avoid such ill by single-mindedly embracing 'the community' as the national ideal.

Communitarianism, for that is the word for it, urges that people should find their happiness and fulfilment in the larger social body – in the family, the ethnic community and, at the apex, the nation-state.[15] Indeed right from the beginning, Singapore's pursuits for better living have gone hand in hand with an equally urgent 'social project', one that aims to discipline Western-type capitalism by modifying its unfettered freedom and moral callousness. Lee's personal conduct, so full of ethical import, is a perfect reflection of the strenuous effort and sacrifice that have created the modern Asian city-state.

And in Lee's infirmity we see too the magical apparition through which Singapore appears before the world: a one-party-dominated state that delivers employment, quality housing, health care and education to its people; a country with an efficient market economy and a 'Confucian-style' paternalistic rule; an oasis of social and economic stability in a region of political and ethnic tension. Communitarianism arguably helps to achieve all these things. The policy also enables the Singapore State to continuously fashion itself anew. The process gives life to the idea of the State as a moral order or, even more evocatively, a 'moral being' that devotes itself to the care and welfare of the people. Communitarianism allows the State to take on a political ambition for comprehensive rule, as it assumes for itself the role as the final arbitrator of all things in society. Whatever the social consequences, to its conservative admirers this is indeed the magic of the Singapore State: a peaceful, prosperous nation that offers its citizens material security and a community of existential meaning shorn of cankerous, self-seeking individualism.

Magic and mask

To ponder on the nature of the Singapore State is to eventually take us to a singular fact: its social peace and economic success, and the prestige it enjoys among its international admirers, have risen from the ashes of liberal democracy. Certainly Lee and the PAP leaders are reluctant liberals who are quick to amend the social-democratic ideals on which Singapore was founded. Yet for all its tough policies and tight political hand, the PAP State does not make us think of the regime of a Marcos or a Mobutu. On the contrary, for many Western admirers Singapore clearly illustrates the possibility that an independent state freed from European colonialism need not succumb to political chaos, personal corruption and economic decay as witnessed in parts of Asia and Africa. A fine progeny of British rule is found in an Asian postcolonial state of evident prosperity. Even with its liberal failings, Singapore is nonetheless a story of economic and political successes, successes achieved by no less than artfully 'fine-tuning' for its own purposes the Westminster parliamentary system it has inherited.

Thus for thousands of foreign workers Singapore is a Mecca of incomparable wealth, offering employment with wages unmatched by what they receive at home. Having achieved a First World economy and standard of living, the country invites admiration and envy from its neighbours. Internationally, it enjoys a world standing unmatched by its physical size and the volume of its economy. From the Vietnam

War to the volatile environment of post-September 11, Singapore has been the oasis of stability in the region. If only for this reason, few Asian political leaders enjoy greater prestige, just as few Asian statesmen are received more warmly in the White House and Downing Street, than Minister Mentor Lee Kuan Yew, to give his current official title in the cabinet. Nonetheless the magic of the Singapore State has to do not only with its economic success and political stability, but more so with the nature of its political rule over the nearly five decades since self-government in 1959. And the enthralment with which Singapore offers itself is a mixture of wavering apparition and awesome power, an intimate union of appearance and effect, that anthropologist Michael Taussig calls 'State fetishism':

> By State fetishism I mean a certain aura of might as figured by the Leviathan or, in a quite different mode, by Hegel's intricately argued vision of the State as not merely the embodiment of reason, of the Idea, but also as an impressively organic unity, something much greater than the sum of its parts.[16]

The state is a figure of Leviathan might; it is also an idea that captures the ineluctable meanings of moral reasoning and commanding authority. The state is always two-faced. It is an awesome endowment of power brought to bear on its subjects, as much as an abstraction – an elusive 'thingness' with an uncertain 'quality of ghostliness'.[17] The trick is how to separate the two, so that we may confront its real power, yet recognize it as a shadowy being without substance. Taussig approvingly alludes to the thinking of Philip Abrams for whom the state is a mask that both conceals and confronts:

> The state is not the reality which stands behind the mask of political practice. It is itself the mask which prevents our seeing political practice as it is. It is, one could almost say, the mind of a mindless world, the purpose of a purposeless condition. There is a state-system [and] and state-idea.... The state comes into being as a structuration within political practice; it starts its life as an implicit construct; it is then reified...and acquires an overt symbolic identity progressively divorced from practice as an illusionary account of practice.[18]

State fetishism then must stage the realities of power and legitimate violence on the stage of a spectacular legerdemain. As such the idea of the state has to be taken seriously, but we must also in a sense not believe in it, not give in to the allurement of its (empty) existence:

> [We] should recognize that cogency of the *idea* of the state as an ideological power and treat that as a compelling object of analysis. But the very reason that requires us to do that is not to *believe* in the idea of the state, not to concede, even as an abstract formal-object, the existence of the state.[19]

For the state, the practicalities of power must need disingenuous 'masking' so that they come across as something else, as tools for realizing its deep commitment to the well-being of the people, as the point or symbol of national identity, as a mirror of the moral credentials of the leaders, and more. Involved in all the things the state does and says, this 'masking' is an important source of its mesmerizing appeal. The mask conceals the state and brings it to 'life'; it is like the one Jim Carrey puts on in his 1994 film before his nightly prowl, transforming him from a lifeless nerd to a smooth-talking, sure-footed dandy – a mask that is seared on to his face.

This book is about the elaborate legerdemain of the Singapore State. With the major public media under its control, the State is ever present on prime-time television and in the papers, explaining the one thousand and one things that it does and proposes to do, and cajoling people into accepting them. A garb of reason and solicitude is draped on the busy, aggressive posturing. Nonetheless it would be wrong to think of the State's action on these occasions as merely for creating, in Abrams's phrase, the 'triumph of concealment'. From the fight against the morally polluting 'yellow culture' to the judicial ruling on oral sex and the caning of an American teenager for vandalism charges, these events are remarkable for the spellbinding ideas they are made to carry. However improbable, and even as people sometimes see through them, these ideas embellish the appearance of the State before the world. The state-idea and the state-effects join in perfect union; each gives life to the other. Not only the opening of a community centre or the Prime Minister's visit to the housing estates, but the caning of the socially recalcitrant, for example, is 'for the public eye' even when held behind the prison wall. For painful punishment of criminals is not only to warn and deter; it is also to circulate important ideas about the State: its duty of care and its unswerving will to incarcerate the socially irresponsible. That is why the State's action, in this and other instances, tends to take on something of a public drama. And the public drama stages the august performance of the

State as all grit and resolve, indulging in no liberal sophistry when it comes to protecting society from those out to destroy peace and order. Crimes pose a threat to society; they are also opportunities for the State to 'display' its will to punish, and the means of legitimate violence at its disposal. Perhaps for this reason, some criminal cases like those dealing with national security are given wide publicity; they are eagerly seized upon by the State to tell 'the official view' on these matters. With these undertakings, the State continuously refashions itself as an idea, and as a 'thing' with a visage of terrifying power and moral authority.

State theory

In modern European thinking, the most important formal feature of the state is its 'public power'. A state has an identifiable geographical territory and within its boundaries it enjoys absolute dominance over all other associations and social groupings. It controls the resources – human and natural – and instruments of force; it claims 'sole *imperium*' without rivalry in the territory.[20] In most versions of state theory, the state is the source of law. But the state is itself based on something beyond itself, on 'rules which have some degree of universal recognition within the territory'.[21] These rules thus support and yet restrict state power. The state is, in somewhat contradictory fashion, 'a legally circumscribed structure of power of supreme jurisdiction'.[22] In modern European political philosophy, the main tension in the approach to state power has been about how to reconcile the fact that the state makes law with being itself subject to what the law sets out. Its 'sovereign jurisdiction' and the way to rein it in both have to be recognized. Much of this way of thinking about the state can be traced back to the seventeenth-century English philosopher Thomas Hobbes. For Hobbes, since man is 'in nature' selfish and contentious it requires an all-powerful state to create peace and order by way of disciplining men's unruly passions. When he declares that 'men are by nature equal', and therefore freedom should be accorded to them, he never loses sight of their wild inclinations and propensity to bear over their fellows. Men 'naturally love liberty [and] domination over others', he writes.[23] In order to keep peace, people have to 'reduce all their wills' to 'one will' (of the state); this is done with popular consent, however. In a political commonwealth, Hobbes asserts, 'men agree among themselves, to submit to one man, or an assembly of men, voluntarily, on confidence to be protected by him against all others'.[24] Legitimacy of rule is important to the all-powerful Leviathan. The state enjoys its power only because men transfer their 'natural rights' to it, and also because it keeps its 'promises and covenants'. For all its might, state power is thus 'limited' because it is conferred by the people who mutually agree to give up, for the public good, the right to govern themselves.

Hobbes was among the first European political philosophers to think about the state's 'public power' in this double sense. After Hobbes, we begin to think of the state as regulated by rules, and the formal power it enjoys must be tied to the people and specific ends of maintaining order and defence. The state makes law, but there are other sources of law, such as customs and traditions. 'The whole point of the natural law approach is to argue that the state is subject to law or that "law predates the state".'[25] Hobbes's 'natural law' posits an idea of man as naturally 'against society', but it is possible to perceive of nature in more benign terms. For his contemporary John Locke (1632–1704) the state of nature is a state of liberty. In the natural state individuals are free 'to order their Actions, and dispose of their Possessions, and Persons as they think fit, within the bounds of the Law of Nature, without asking leave, or depending on the will of any other Man', as he writes in *Two Treatises of Government*.[26] 'Good government' is 'government by consent' and one that recognizes the naturalness of liberty and justice as God-given.

Here Locke reveals something of the difficulties in the thinking about state power in general. All state theories have to deal with how to make state power something that can extract compliance from people – naturally and rationally rather by the instruments of force at the state's disposal. This is the question of legitimacy, of the right of the state to 'will' its will on the people. All the talk of 'the state of nature' is really about trying to find the 'absolute cause', the final reasoning, that gives state power its particular form and potency in relation to the people and society. Hobbes and Locke shared the view that it is the 'social contract' that gives the state its rightful status and power. For Locke in particular, 'government by consent' enjoys legitimacy because it gives heed to the liberty people enjoy *naturally*. The state must recognize a higher order than itself: the 'state of natural liberty'.

State and moral authority

This is the compelling heritage of modern state theory. In our own time, the state's public power and the 'higher authority' on which it is based must be something from the popular will enshrined in legal institutional forms. Constitutional government protects the people of their freedom and properties, and from each other as from the state. The state is anxious to ensure the continuance of its power; it is even more anxious to show that it acts with reason, and in the interests of the people. For any state regime, the more absolute its hold on society, and the more assured it is of its 'sacred status', the busier it is in telling the people what it is trying to do. The propaganda machine is the most organized in the authoritarian state. State power is never 'naked'. Even under a Hitler or a Stalin or a Mao, the state still needs to find legitimacy for its power and right of existence. In communism we have that peculiar, unapologetic phrase 'dictatorship of the proletariat'. On the other end of the political spectrum, in modern democracy 'supreme legitimacy' is found in the mandate of popular elections. These forms of state power are of different ideological and structural kinds, yet they have the same roots in 'social contract' and 'government by consent', despite the twisted meanings regimes have given these phrases.

'Dictatorship of the proletariat' and popular electorate mandate, in different ways, provide the modern state with the source of its authority. And this authority too has to be cloaked in the legerdemain of potency and reason. What makes the state even more enthralling is the way it takes on a moral stature before which we bow and 'bend our will'.[27] Of course, all forms of authorities demand this from us, and their rules and commands impress upon us that we are not free to do as we please. But state authority based on moral stature would do this in a quite different way. When we give ourselves over to the care of the doctor and swallow the bitter medicine he prescribes, we do so willingly because it is good for us. When a prisoner obeys the commands of the guards, it is to avoid the pain of punishment dished out to those who defy the rules. In both cases we yield to authority because the act has a practical effect: it brings us benefits (in the form of health) or it avoids punishment.

However, when we obey someone with moral authority – say, a religious leader – we do so out of our own will. Moral authority exhorts compliance from us not by the enticement of stick and carrots; it does so from its inherent 'rightness', and because the person carrying it has a particular 'ethical excellence' that makes meaningful the word 'moral'. Morality, Emile Durkheim writes,

> constitutes a category of rules where the idea of authority plays an absolutely preponderant role.... It is a certain and incontestable fact that an act is not moral, even when it is in substantial agreement with the rule, if the consideration of consequences

has determined it. [F]or the act to be everything it should be, for the rule to be obeyed as it ought to be, it is necessary for us to yield, not in order to avoid disagreeable results or some moral or material punishment, or to obtain a certain reward; but very simply because we must, regardless of the consequences our conduct may have for us.[28]

Moral authority, we may say, garbs the state with the most magical of forms and effects. When ably handled by the 'ministry of culture', it extracts compliance from the people who yield because they wish to. Here we begin to see the rich significance when we speak of 'the Father of the Nation' or remember the heroic struggle of those who brought the nation – and thus ourselves as national subjects – into being. The state as moral order is state power at its most subtle and magical. Better still is to give the state a personality culled from the values and individualities of its leaders; the state as moral being takes on its duties, not for its own continuance, but for fulfilling its highest calling of selfless devotion to the people. Moral authority, when the state achieves it, delivers two things of central importance. It nurtures a special force that makes us naturally surrender our will. And it creates the appearance of moral solicitude that overlays the reality of 'heartless bureaucracy' and judicial violence that are also central features of the state.

'Supply-side socialism'

Moral authority of the state is important here because it is also a fond phrase from the lips of Singapore's leaders. Singaporeans will no doubt remember the so-called Catherine Lim affair. In November 1994, the novelist Lim wrote in the *Straits Times* pointing to the continuing strong hand of the government under the then new Prime Minister Goh Chok Tong.[29] Goh had just taken over from Lee Kuan Yew after the 1991 general election, and it was eagerly speculated whether he could strike out in his own direction or would follow Lee, then holding the influential position of Senior Minister in the Prime Minister's Department. The Prime Minister had read Lim's comments as saying that he was not his own man, and accused her of undermining his 'moral authority' and thus his government's effectiveness to rule. In the words of his press secretary, the government did not allow 'journalists, novelists, short-story writers or theatre groups to set the political agenda outside the political realm'.[30] In a kind of ritual public humiliation, Lim responded to the PM's rebuttal by withdrawing her statements. For such a matter could not be the subject of equal exchange between citizen and elected government, and Singaporeans quickly recognized how seriously the State would defend its 'moral stature'. As always Lee, who holds on to an idea to the point of obsession, would set an example. His public pronouncements are littered with references to the necessary honesty and integrity of people in the government, qualities which he himself best exemplifies. PAP leaders must not only show devotion to public duties and quickness of response to real or imaginary crisis, but they must also be unfalteringly moral in personal life and honest in financial dealings. Often extra-marital affairs are a part of the investigation of those the State seeks to question, and Lee himself has come down hard on those in the PAP inner circle thought to have failed by his standards.[31] It is hard to imagine having a Chris Carter, the first openly gay member of the New Zealand parliament, or a Jeffery Archer in the Singapore government.

The ethical standard of the government and its leaders reflects the larger enterprise of the State of making itself a 'being' of principle and moral reasoning. With Singapore's collective welfare in mind, the State takes up, with stark simplicity, the idea of 'the community' as a way of galvanizing the nation under the PAP. The State sponsorship of the idea produces some interesting results.

For all its uneasiness with welfare handouts, Singapore has in fact a sophisticated system of delivering social and financial help to the needy. It does not however take the form of direct payment to individuals. Instead the State actively supports voluntary associations and the ethnic community organizations, and financial subsidies are given to them to be redistributed as they see fit. The State is prepared to pay up to 80 per cent of the operating costs of the self-help groups. The system thus aims to do several things: it delegates some of the government responsibilities to the communities; it helps to ensure their viability as meaningful institutions to which people would like to belong; and it curtails individual dependence on the government.

The other notable feature is the way funds are distributed along eth-nic or racial lines. Currently the three major community organizations are the Chinese Development Assistance Council (CDAC), Yayasan Mendaki and the Indian Development Agency (IDA) for Chinese, Malays and Indians respectively. In official terms, these organizations help children of working-class families to do better in schools; most of the monies they receive go toward providing tutors and small con-tributions for books and stationery for students. Clearly government support is meant to be a kind of financial and morale bolstering for each community. The CDAC, for example, has 2,000 volunteers who carry out the various self-help schemes. In 2002 it had spent only 7 cents of each dollar on administration, and the charity dinner that year raised $1.65 million from various foundations, clan associations, commercial companies and individuals.

In helping to build up each organization, the government channels money into its coffers through a complex system of financial subsidies, while leaving it to run much of its day-to-day affairs. It has put in place a law of compulsory contribution by all citizens of each ethnic group. Thus the Chinese contribute to the CDAC, Malays to the Yayasan Mendaki and Indians to the IDA. To implement the law, the government collects on behalf of these organizations by deducting from an employee's monthly superannuation saving, the Central Provident Fund (CPF). Unless a person writes to state his or her wish not to contribute, the deduction is mandatory. Since the amount is small – actually a few dollars a month – very few people opt out. In this way, a regular flow of funds is collected and then redistributed to each community organization to finance its various education and skill improvement programmes.[32]

Now, the government could have run these programmes directly and funded them from general taxes. Besides, the scheme tends to invite the criticism that it works against 'national integration' by playing up ethnic and communal distinctions. The government is itself keen to avoid 'the undesirable scenario where more and more social services and programmes are organized along ethnic lines'. 'After September 11', the then Prime Minister Goh Chok Tong said at a CDAC annual dinner, 'we should be even more alert to any policy or programme that may accentuate racial and religious differences.' However, he did not believe that the self-help groups had sharpened the 'racial divide'. As to the suggestion that the government should take the community programmes into its own hands, the Prime Minister's position is that 'Trying to solve individual social problems through centralized government departments will not be as effective as the community taking an interest in them'; the community is simply better at giving 'warm, emotional support' than 'impersonal, efficient bureaucracy'.[33]

Renowned for its direct intervention in much of the economy and social life, the State seems to be acting out of character. But the Prime Minister clearly recognized the alienating effects of bureaucracy, and in his view creating communal cohesion along ethnic lines is worth the risk of possible ethnic and religious divide – a volatile subject in 'multiracial' Singapore. Nevertheless the community self-help scheme gels with State design in some important ways. For one thing, the

scheme accords with the State's virulent opposition to social welfarism, something it regards with an almost apocalyptic distaste, while
continuing to stick to some form of the 'socialist principles' of the
past. Direct welfare handouts are, in its view, a policy of multiple ills.
It leads to budgetary blowout, raises labour costs, and saps personal
initiative and communal responsibilities. State welfarism brings the
serious social, economic and moral disasters that so evidently plague
the West, the people are frequently told. Leaving aside the historical
reason for the moment, once the State has decided that it will adhere
to some of its 'social-democratic' principles, it must then attack the
ills of welfarism with rigour.

'Market socialism'

In Singapore, state delivery of social goods has been called 'supply-side
socialism'. This form of 'socialism' is to be made distinct from Soviet-
style socialism and the crisis-ridden welfare system of the Western
liberal states; it does not encourage sloth and dependence. It works
instead to 'maximize the ability of all human beings' and 'help to
make the national economy more competitive'.[34] As a local sociologist
explains,

> [The Singapore State is keen to contrast its policy with] any form of
> direct cash transfer from the state to individual citizens – what is in
> Singapore popularly perceived as 'hand-outs' by liberal 'Western'
> welfare state which supposedly saps the work ethic. The idea of
> 'supply-side' socialism is reflected in the redistributive processes
> in provision of public housing to help maintain security and
> stability of households, in the logic of education as human capital
> investment, in the provision of superior infrastructure as input
> in efficiency and market competitiveness and in the substantial
> subsidies provided for the voluntary welfare organisations which
> assist the various constituencies of socially disadvantaged.[35]

The State clearly has in mind the conventional 'demand-side socialism'
where citizens can expect as a *right* the state's obligation to pay up.
'Supply-side socialism' is meant to distance itself from the practices
of the Soviet-style economy with its poor market mechanism and
the massive and inefficient bureaucracy that ran it, and makes a
stand against the self-interest-driven, competitively violent, liberal
capitalism. In rejecting these systems, 'supply-side socialism' can have
the best of both worlds.

For one thing, it sees welfare assistance, public housing, health care and so on very much in terms of 'human resources development'. Social redistribution must remove people from the labour market. Instead 'supply-side socialism' subjects people to the strenuous logic of the market which, like the Darwinian survival of the fittest, rewards the disciplined, the adaptable and the intelligent. At the same time, the forms of state delivery also take some of the sting out of the brutal competitiveness of the capitalist system that casts many people by the wayside. In Singapore, capitalism, to which it is deeply committed, is meant to have a human face, one that is endowed with a certain moral discernment. The taming of capitalism, in this sense, greatly concerns the Singapore State. Indeed, we recall, the official support of ethnic communities is meant to create cultural belonging and is thus a panacea for the heartlessness of market capitalism. Nonetheless each ethnic community, for all the talk of cultural cohesion and belonging, must still find its place in the economy. An ethnic community falls short of the ideal unless people are employed and participate in the economy. In the State's vision, 'community' is the stuff of existential meaning and moral consideration; it is also a template of how people can have a useful role in the world.

'Socialism that works'

In making ethnic organizations culturally and financially viable, the State harvests for its own purposes the moral force and social co-hesion they seemingly hold. The ethnic communities become the 'building blocks' of the nation. They are, like the nation, a subtle interplay of material enjoyment and existential wholeness. Imperceptibly the nation becomes a community writ large. The prudent 'socialist programme' without budgetary blowouts; a capitalist economy that avoids the culture of 'everyone for him/herself': these are the stuff of nation-building. At the most ideal, Singapore will do away with the 'social contract' and the bargaining of citizenship rights and obligations. There is instead a 'national community' of Asian cultural values and postcolonial certitude under a guardianship of the PAP. The 'community' and 'socialism', with the meaning they have come to have in Singapore, are very much the pillars of State power.

For its varied ideological shades, 'socialism' is not a strange term to be spoken of in Singapore. Historically its roots were imported from the post-war British Labour Party by the first-generation leaders like Lee Kuan Yew, Goh Keng Swee and Toh Chin Chye who were students in England at the time. Goh, the first Financial Minister and architect of the island's industrial development, attended the London School of Economics in the early 1950s; and he has explained the government's 'socialist' visions this way:

> We tried to achieve a society where all citizens could have a decent living.... Taking an overall view of Singapore's economic policy, we can see how radically different from the *laissez-faire* policies of the colonial era.... These *had* led Singapore to a dead end, with little economic growth, massive unemployment, wretched housing, and inadequate education.

In contrast, in order to achieve a better standard of living for the people the Singapore State would be much more directly involved in the economy:

> We had to try a more activist and interventionist approach. Demo-cratic socialist economic policies ranged from direct participation in industry to the supply of infrastructure facilities by statutory authorities, and to laying down clear guidelines to the private sector as to what they could and should do. The successful im-plementation of these policies depended on their acceptance by the people, generally, and on the active cooperation of organised labour in particular.[36]

'The welfare state' is the term we use to describe the social policy of the post-war West. It is concerned with matters of social equality and redistribution; and state revenue that funds these comes largely from taxes. After the rapid privatization in the 1970s the state is increasingly divorcing itself from economic intervention and ownership in carrying out its social welfare obligations.

Singapore took a different route, and the approach is in spirit that of an 'economic state'. Here the principle is for the state

> to promote the well-being of the vast majority of the people by administering and boosting economic production, equalising resource distribution, and guaranteeing work opportunity and economic security. This in theory would eradicate the source of widespread poverty and inequality.[37]

'The economic state' funds its social deliveries by taking a direct hand in the production of goods and services. In this, Singapore did not follow the path of the nationalization of key industries as in post-war

Britain under Labour. Its strong economic hand is visible instead in a system of direct or indirect ownership. Since independence it has set up various statutory bodies that own and run companies and infrastructural services free from direct government intervention; the Housing Development Board (HDB) which builds and sells to the public affordable housing is a good example. Some of these companies are solely owned by the State, or partly or totally privatized where the State is the majority shareholder. Through such arrangements, the State has a huge stake in major industries like shipyards, defence equipment, the media and telecommunication. Having trusted people on the management boards, the companies are to be run on commercial principles. They are expected to be efficient and make profits. They are not to be given special privileges or allowed to conceal any subsidies they receive.[38] It is hard to estimate how much each year these companies bring to the state coffers. Many of the government-linked companies (GLCs) come under Temasek Holdings with a global portfolio of $103 billion, mainly in Singapore, Asia and the OECD countries. Temasek is the largest commercial corporation in Singapore, whose CEO (chief executive officer) and executive director is Ho Ching, the wife of the Prime Minister, Lee Hsien Loong.

The GLCs have helped to fund Singapore's 'supply-side socialism'. They are important not only for the money they make, but also for the employment and good economic climate they provide. For its kind of social welfare system, the PAP had preferred the ideologically more subtle 'socialism that works',[39] with its suggestion of a state economic role, tight regulation of the labour movement, and provision of public goods like housing, health and education. History has no small part in this. If PAP elders like Lee Kuan Yew and Goh Keng Swee were men of vision who tried to realize the agendas of Fabian socialism on the tropical island, it is also true that they saw the political necessity of it. Indeed, conditions in the nation's early years gave 'democratic-socialism' a strong pragmatic, economic slant. In 1963 President Sukarno's *Konfrontasi* to subvert the newly formed Malaysian Federation stopped all entrepôt trade with Indonesia, and later in 1965 the expulsion from the Malaysian Federation destroyed the dream of a Malaysian common market for Singapore goods. In July 1967 the British announced their intention to withdraw their military forces in Singapore. The news threatened investment confidence, and the withdrawal would do away with some 45,000 jobs at the British military bases on the island. Not only the welfare of the people but Singapore's very economic survival was at stake; lifting the nation from the impending doom became a critical task.

As policies were put in place to do this, they also gave the State a new lease of life. Over the following decades, as Singapore faced other crises, from the 1987 recession to the current 'war on terror', each was an opportunity to show how skilfully the State had met the problem head on, each an occasion to show how energetically it would avert the nation from disaster. In any event, there is no getting away from the fact that the 'moral authority' of the State and its leaders rests on their economic quests, on their successful delivering of jobs and better living. But what kind of men – and, less often, women – are these to whom Singaporeans are asked to look up with awe and admiration? These are, in the first place, people of technical and managerial skills. As for the PAP elders who led the nation to independence, their political skill and fiery passion are almost redundant in the present endeavours:

> The first-generation leaders are the men who led their people to independence. They seldom understand that government means more than just mobilising mass support for protest against the injustices of colonialism, so, after independence they cannot deliver the goods. They had not learned about administration and economic growth. They are therefore not able to create confidence in their government's promises and undertakings: they cannot get foreign investment to add to their domestic capital; and they have not educated and trained their young in the skills and disciplines which could use this capital and the political machinery to bring about a better life.[40]

Lee is here pronouncing his famous pragmatism. Charismatic personalities and the ability to work the crowd are all very well, but they have little use in improving the traffic or running the economy or attracting foreign investment. When it comes to that, in contemporary Singapore people become political leaders primarily by being good at these practical enterprises. In this context the ideas of 'personal attributes' and 'moral authority' appear to take a back seat; the trick is how to convince the people that these are still important, and that they still endow the managers and technocrats who make up the new generation of leaders.

Shared Values

Investing moral authority in technocrats is of course a serious contradiction. If 'moral authority' is to mean anything, we recall, it is because the words and acts of great leaders demand respect for what they are. Precepts of moral import must be obeyed 'out of [our] respect for it and for it alone'.[41] When we follow our leaders because we fear them, or because we want what they can do for us, then we are not far away from being led by our selfish instincts, unmoved by the grander purposes the distinguished figures embody. For Singaporeans to respect the Singapore State for what it has done for them is a bit like saying Catholics respect the Pope because he has balanced the Vatican budget or achieved a quantum leap in annual conversion to the faith. For the Singapore State the issue is: with much of the influence built upon its 'supply-side socialism', it is never sure if popular compliance is due to its social deliveries or to the PAP's moral credentials. Most liberal regimes would accept that political rule is a bit of both; victory at the ballot box is decided as much by policies as by the dynamic appeal of the party leaders. But this is not a happy option for Singapore.

And it says much about the State's hegemonic ambitions that it wants 'morality' to be the primary basis of its authority and the reason for its popular support. The State is not blind to the Durkheimian insight. If the PAP State can be perceived as driven by the highest, most ethical of motives, then people will accept its prestige and influences *for themselves*. Popular support is not something so crass as to relate to what it has actually done for the people. In the high-mindedness of PAP leaders, for the people to give allegiance to the State only because of what it will do for them sounds like 'bargaining'; it is like his disciples telling Jesus that they would believe in him only if he would heal another sick person or perform another miracle.

Embodied by men of honour and selflessness, the State must transform popular compliance into faith, always blind to a degree. The result is to put State power on surer ground. If it can be nurtured among the people, popular faith in the State creates tolerance for changing policies and harsh state measures. With popular faith, people will not be so quick to take the State to task at the first sign of a broken promise. Rule by 'moral authority' removes State influence and power from the fluctuations of electoral politics: this is the PAP's most elegiac ambition.

Electoral politics is a tricky thing for the PAP. It has often complained that Singaporeans tend to vote with their feet. This is a complaint about the fickle-mindedness of those who either do not turn up at the poll (Singapore has no compulsory voting) or cast protest votes in favour of opposition parties. In Singapore general elections are not for deciding which party should govern, which in any case is never in doubt, but an occasion for the long-reigning PAP to reaffirm its political stature and legitimacy. PAP candidates enter elections with a strong sense of mission and the sweet security of success. That is why, when the party loses one or two seats, it goes into profound soul-searching, and younger leaders begin to think that they are not worthy of the trust the party elders have placed in them. Accustomed to winning all the parliamentary seats, the PAP sees something darkly foreboding: the erosion of the moral status that it too confidently assumes for itself. To use the ballot box for casting protest votes and for getting better services in the housing estates will not do. Worse still, it goes against the ethics of 'community' enshrined in the nation, now perilously betrayed by the formalities of democracy.

The idea of 'community' is crucial for the State because it is meant to tie up the many loose ends of political rule by 'moral authority'. To be sure, legitimacy and social support would come from welfare delivery and efficient administration. Nevertheless good government must also personify the values of sacrifice and incorruptibility of the PAP leaders. In any case, once the State has attained a reputation of excellence then it can expect to command popular support by its moral status – quite apart from its good policies. Instilling the right values in the people and investing in the State a distinct moral stature are better ways of ensuring political power, for people in a 'community' would yield to the larger social collective and voluntarily limit their own needs and demands. As an ideal, 'community' can be all these things: a social grouping of existential and cultural meaning; an association of people bound by cooperation and consensus; and a 'total society' where 'all is one', where the state is the major arbiter of things and a focus of individual loyalty and aspirations. That is what 'community' looks like when it becomes a political ideology.

In January 1991 the PAP government laid before the parliament the Command Paper *Shared Values*. The then Prime Minister Goh Chok Tong had earlier suggested the need for a national ideology on which 'to anchor a Singaporean identity'. The national ideology is to include a set of values which 'Singaporeans of all races and faiths can subscribe to and live by', incorporating 'our heritages, and attitudes and values which have helped us to survive and succeed as a nation'. The Command Paper lists the five Shared Core Values as: 'Nation before community and society above self', 'Family as the basic unit of society', 'Regard and community support for the individual', 'Consensus instead of contention' and 'Racial and religious harmony'.[42]

With *Shared Values* the State very much bares its soul. In a way that would worry the progressive liberals, there is no mentioning of personal liberty, civil society, and constitutional restraint of governmental power. Instead greatest weight is put on 'society' and 'the family'. 'Society above self' and 'family as the basic building block of society' both clearly inscribe a position for individuals who must, as a matter of ethical choice, temper their personal interest and consider the larger social good. The position is intrinsically 'right'; it is also time-tested to be of practical use. 'Putting the interests of society as a whole ahead of individual interests has been a major factor in Singapore's success', the Command Paper declares; it goes on to say:

> This attitude has enabled the country to overcome difficult challenges, such as the withdrawal of British forces in the early 1970s and the severe economic recession in 1985. If Singaporeans had insisted on their individual rights and prerogatives, and refused to compromise these for the greater interests of the nation, they would have restricted the options available for solving these problems. Promising solutions would have been ruled out as unacceptable to one group or another. Instead, Singaporeans have shown themselves willing to make temporary individual sacrifices for the sake of the group. The results have shown that over the long term, this leads to greater success for all.

At the end, the buttery wheedling of *Shared Values* sounds like an admission that something is amiss. To the PAP State, people have often forgotten that the goods of 'supply-side socialism' emanate from the 'inner goodness' of those in power; they are not 'bribes' to purchase electoral votes. Just as the PAP leaders are moved to serve as unfailingly as the arrival of the monsoon rain each November, so each citizen's abiding to the 'national agenda' has to come from deeply felt instinct. Like the leaders, the people of Singapore must go not for the crass calculation of 'more for me and less for you' but for total support for the heart-warming 'national community'.

This of course is ethical impulse at its most organic and intuitive. It is hard not to read *Shared Values* as laying out the ways of achieving this disciplining of the Self. That PAP leaders like Lee so evidently practise this, that PAP is the emblem of it, raises the State like a ghost of disquieting presence. The best is to instil personal sacrifice and regard for society as 'values'. Then the people will instinctively feel the ethical worth of these 'values' and faithfully tie them to the PAP's grand political project. 'National consensus' will be achieved without the State having to yield its iron hand. Public administration and delivery of social goods are, as in any society, part of the State's public duty. However, in Singapore these enterprises must not obscure the State's lofty purposes and its almost sacred stature.

And in all this *Shared Values* turns to the family like some talisman of magic. 'Society above self' in fact sets up a ladder of increasing demands and responsibility: the family is the foundation on the apex of which sits the monumental artifice of the nation-state:

> The family is the fundamental building block out of which larger social structures can be stably constructed. It is the group within which human beings most naturally express their love for parents, spouse and children, and find happiness and fulfilment. It is the best way human societies have found to provide children a secure and nurturing environment in which to grow up, to pass on the society's store of wisdom and experience from generation to generation, and to look after the needs of the elderly.

In the State's vision, the family is best for bringing moral sensibility to people's lives because it is the most social: we all come from families; we all have a mother and a father. The social is moral because it is a way of relating to other people, of living in solidarity with a group, as Durkheim says.[43] Community and the nation are often too abstract and general, but the family is filled with the stuff of blood, loyalty and sentiment. A reservoir of social values, the family is simply that place in which love, feelings and other virtues brilliantly dwell, offering as it does nurture and fulfilment for its members. Some might recognize the Victorian view of the family in *Shared Values'* celebration of homely decencies:

> In recent decades many developed societies have witnessed a trend towards heavier reliance on the state to take care of the aged, and more permissive social mores, such as increasing acceptance of 'alternative lifestyles', casual sexual relationships and single parenthood. The result has been to weaken the family unit. Singapore should not follow these untested fashions uncritically.

People brought up in good families, we are asked to believe, are socially responsible and sexually disciplined; they take care of the young and the aged, enjoy sex within marriage, and make the best citizens. In the family the individuals are shaped into moral beings who tie their personal happiness to the nation and see the purposes of the State as naturally echoing their own.

Epilogue: dangers of bar-top dancing

In Singapore as in other cities, patrons of dance clubs are an exuberant lot. Young women seem especially keen to get into the swing of things. In inebriated high spirits, they are likely to get up on the bar to execute a few bum-wriggling, breast-shaking moves. Liberated types as they are in Singapore, and after not a few Bacardi and Cokes or Tiger Beers, they forgive the male leering from the floor as their bodies whirl and short skirts fly. In intoxicated abandonment, they throw caution to the wind; perhaps the risk of falling and inviting less than gentlemanly stares from the men looking up their skirts are part of the thrill.

In October 2002, bar-top dancing entered parliamentary debate when Minister of State Vivian Balakrishnan raised the subject on the floor. The Minister is certain of the dangers:

> If you want to dance on the bar top, some of us will fall off that bar top. Some will die as a result. Usually it is a girl with a short skirt who's dancing on it, who may attract some insults from other men. The boyfriend starts fighting. Some people will die. Blood will be shed for liberalising the policy.[44]

Bar-top dancing is perhaps like adventurous play in kindergarten: one risks a bloody nose or a bruised knee for the fun of it. And similarly there is a price to be paid when the State eases its rule, so the Minister warns, when it allows people to have more fun and liberty. Still, neither Singaporeans nor late revellers in nightclubs need reminding. When they climb on top of the bar, the 'risks' are simply the nature of it. When some mishaps occur people can handle them without fuss and do not need official intervention. In any case, always anxious to protect public safety the government quickly put a ban on bar-top dancing, imposing a fine of up to $10,000 on nightclubs and bars for breaching licensing laws. The ban was a disaster for the tourist industry struggling to lure visitors back after the SARS epidemic. Besides it was a time when Singaporeans were being cajoled into making a 'shift in mindset' by trying 'greater risk-taking, experimentation, diversity, choices and decision-making', as the then Prime Minister Goh Chok Tong told the nation.[45] So he reconsidered; maybe Singapore could use a few 'bohemias' (*sic*) where people could gather and be creative. 'Studies in the U.S. have shown that entrepreneurship is closely correlated with the level of cultural vibrancy', he explained, and 'creative folk needed a thriving music and arts scene, openness and diversity'.[46] In August 2003, the ban on bar-top dancing was lifted but with some provisions. Restrictions are now placed on paid performers, who must not 'mingle with customers'; there is to be 'no chatting and/or drinking with patrons'. The police will not enforce the 'no-mingling rule' if exemption is sought from the Singapore Tourism Board, which will make the decision after considering such factors as 'the nature of performance' and 'the perceived tourism value of the outlet or event in contributing to the overall attractiveness of Singapore's night scene'.[47]

The regulation of bar-top dancing reflects the State's public face. The recklessness, the leery looking up of skirts, and fights between jealous males are, in the way the Minister of State imagines them, lascivious and dangerous. The provision of 'no mingling' between paid performers and customers is a kind of halfway house in which the State can relax the rule to a degree and still put a lid on the things it disapproves of. Indeed this has become something of a pattern in the current 'softer PAP rule'. When Goh announced that he would now allow bungee-jumping and bar-top dancing, he made the point that 'it is not because I encourage these activities'. Similarly, although the public service now employs gay men, the State makes clear that it does not encourage or endorse gay lifestyles, and it has kept the law against oral and anal sex on the books. All this is to remind the public that the State can, when it wishes, turn to its splenetic side. In loosening up rules and regulations, the State never gives up the near-apocalyptic vision of economic doom and social collapse. The lesson may well be this: greater liberty is not only the intoxicating stuff that frees the body and fires up the imagination, but something of moral and physical danger as well.

In one sense the State is merely carrying on like the best of the first-generation leaders, whose anxiety and apprehensions are but signs of virtue. Lee had first conjured up Singapore as a nation with a sacred mission: to give its people the material life that colonialism denied them, and to make 'democratic socialism' work by changing some of its rules. And Lee sick from worrying is the best metaphor of the deep moralism of the State. The National Father fallen sick set the example of how deep moral ethos could turn weaknesses around, and remind people of the need for sacrifice and hard work in order to prevent the nation from economic lethargy and social chaos. However, over time this moralism also hardened into something else at once violent and intolerant of signs of social recalcitrance.

In Singapore the State's moralism has given rise to an efficient government and a modern social vision; but it also produces hysterical self-righteousness and thus fear of otherness. Mixing high moral principle with fear of strangers has bequeathed Singapore with an indignation that quickly identifies rivals and 'evil others' of different sorts. Their suppression by the State always seems like a moral panic. Singapore's social and economic success has allowed the PAP to claim, so frequently as to be self-justifying, that it has the only viable political vision for the country. What is this except a moralism, as I have called it, that knows with unwavering certainty what is right and what is wrong? Drawing a clean line between them assures the State of its unfaltering position, and helps to seek out, and eliminate, otherness and its harm. Just as the Puritan Fathers of America found it easy to make out their satanic foes – witches, Indians, Quakers – because God was on their side, so does the Singapore State confront its otherness with the security that it is the only right thing for Singapore.

And there is no question of keeping this to itself. With so much at stake, not least Singapore's continuing prosperity, the restless energy and feverish imagination have to be passed on to the people and made into national culture, a culture akin to trauma. The State's 'worst scenario' approach cuts deep in *Shared Values*. It is not enough that people accept 'society above self' and so on; they must be made to feel deeply the urgency of what the State is trying to achieve so that they too are constantly alert to Singapore's uncertain future. The awareness – that Singapore's good life cannot be taken for granted – is evident in the Shenton Way business district, in the humming of the factories and the official announcements which mix local achievements with warning of the dark cloud over the New York Stock Exchange.

If the state is the actuality of an ethical idea, a social collectivity in which an individual can achieve selfhood, what then is at the centre of the Singapore national subject? What lies there is surely this driving impulse: not only must he work forever harder and more efficiently in the competitive economy, but he must also take upon himself the gnawing anxiety of the State. *Shared Values* with its emphasis on national consensus is meant to make this demand on the people. For a national subject to be as obsessive as the State with Singapore's real or imaginary dangers is merely being a good citizen. Singapore's national culture is built on this 'culture of excess' forever making people sit on the edge of their seat, ever ready to take action to avert the nation from the fate of social and economic collapse. Anxiety is an effect of impending danger, even though such danger may be largely imagined, as Freud has said.[48] Nonetheless, anxiety is experienced as real, and the subject responds in a definite way. And in this response, anxiety begins to take on a positive quality in preparing the subject against fright. 'There is something about anxiety that protects its subject against fright and so against fright-neurosis', Freud writes.[49]

We can think of Singapore's national culture in much the same terms. The national imagining of doom, so arduously nurtured by the State, is at times debilitating and yet seemingly incites the nation to quickly respond to crisis. And this rallying behind the flag is not only due to the strong hands of the State. For all its totalizing political ambitions, the State is unable to fully carry them out. And the genius of moralism is precisely the way it shows State actions as ethically reasonable and practically necessary in relation to the interests of the people. Singapore is a nation of excess and anxiety. When this is read as efficient government and energetic action, not a few Singaporeans and the State itself would be proud of such a labelling.

2 Trauma and the 'culture of excess'

'[T]he State' never stops talking.
> Philip Corrigan and Derek Sayer, *The Great Arch:*
> *English State Formation as Cultural Revolution.*

The traumatized, we might say, carry an impossible history within them, or they become themselves the symptom of a history that they cannot entirely possess.
> Cathy Caruth, *Trauma: Explorations in Memory*

The State – in excess

When it comes to dealing with political criticism, the Singapore State has always insisted on what it calls 'the need for rigorous and robust response'. But it is often hard to know exactly what form of criticism would invite the State's wrath. For the State does not prosecute all oppositions and people of different political views. Some civil society groups like Think Centre, and AWARE for advancing women's causes apparently flourish, and books probing PAP's hold on power are available in the bookshops and read in universities. When the State finds it necessary to 'respond', however, it often does so with a heavy hand and high drama.

On 7 October 1994, the Paris-based *International Herald Tribune* published in its opinion page a commentary, 'The Smoke over Parts of Asia obscures some Profound Concern'. The author was US political economist Christopher Lingle on a two-year contract as research fellow at the National University of Singapore, who included in his commentary the following:

> Intolerant regimes in the region reveal considerable ingenuity in suppressing dissent. Some techniques lack finesse: crushing unarmed students with tanks or imprisoning dissidents. Others are more subtle: relying upon a compliant judiciary to bankrupt

opposition politicians, or buying out enough of the opposition to take control "democratically."

Singapore was not mentioned in the article; but it nonetheless felt that the commentary scandalised its judiciary, and attacked its government's prestige and reputation. The police interrogated Lingle. Later he was allowed to leave the country to visit his ailing father in the United States, but did not return. Meanwhile the IHT on being informed of the then Senior Minister Lee Kuan Yew's grave displeasure made an unreserved apology to him and the Singapore judiciary. The government nonetheless pressed ahead with a contempt of court proceeding against the paper. Lee himself sued for libel alleging that the statement could be construed as 'suggesting that he had sought to suppress political activity in Singapore by bankrupting opposition politicians through court actions in which he relies on a compliant judiciary to find in his favour without regard to the merit of the case.'[1] In any case the trial ruled for the government, and the editor, publisher, distributor and printer of the paper were fined sums ranging from $1,500 to $5,000 and ordered to pay for costs of prosecution. The highest penalty of $10,000 was reserved for Lingle. Since he was not in Singapore, his superannuation saving and wages owed to him were duly used to pay the court costs. Faced with the judgement the IHT sought settlement with Lee to whom it paid $300,000 for damages and costs.

For agencies like Amnesty International and International Commission of Jurists the Lingle case is hardly news; it adds to what they have been saying about Singapore's treatment of political dissent. To Justice Michael Kirby of Australia, 'Singapore had not kept up with the developments of the common law in other countries.... punishment for scandalizing the courts hasn't been used in England for 60 years.'[2] For people like Justice Kirby, it is hard to fathom how bringing IHT and Lingle to court would protect Singapore's good name, except by discouraging the more timid commentators from similar ventures.

As usual, the State has its own way of reasoning. The official view has always been that 'politics' is the business of electoral contests; debates about government policies should take place in the proper arena, at the parliamentary floor more exactly. People not elected to the parliament have no right to mess with matters of government. Since foreign correspondents have professional allegiances elsewhere, they cannot be allowed to 'interfere with our domestic politics.'[3] As for the Lingle case, the court's position is that '[if] the criticism impugned the integrity and impartiality of the court, this amounted to contempt of court, even it is was not so intended.' Though the offending passage did not refer to Singapore, it nonetheless raises the question whether a 'reasonable person' would conclude that it did.[4] The judge thought that in the context of the case such person would.

Singapore's tiff with IHT, however, is not a single incident in its relationship with the foreign media; *Far East Economic Review, Asiaweek, The Economist,* and *Asian Wall Street Journal* have in various ways gotten into trouble with the authorities. Like the IHT case, these events are marked by forceful legal argument, rigorous State responses, and punishing consequences for the papers and journalists involved. Writing now, I remember going through the grounds with a government official on some of these cases. Over a cup of Chinese tea gone insipidly cold, with the air-conditioner blowing on my face, I finally got what I could in the way of a firm opinion when he said, 'We need to put a stop to this kind of thing straight away, otherwise everyone will be doing it.' There was no need for me to query: who is this 'everyone'? Looking back, I can be impressed by the richly suggestive statement. Perhaps State action as in the Lingle case posed a kind of pre-emptive strike against what it saw as an assault on the dignity of the State. As a pre-emptive strike, such action would require imagination and tough counter-measure. To do this the State would paint the 'worst scenarios': the loosening up of the press laws would have the foreign media saying all kinds of libellous things about Singapore; allowing greater 'freedom of expression' at home would have the communalists at each other's throat; and allowing mass street gatherings would have the radicals and political opportunists rousing people's irrational passion and leading them to riots. The logic of the IHT-Lingle case is the logic of 'over-perceiving' the harm on the State and the judiciary. If 'everyone will be doing it' sounds like an impatient quickness of the tongue in an interview seemingly going nowhere, the phrase nonetheless opens up to the expansive horizon in which all manner of people – some real, most imaginary – are trying to bring chaos and disorder to Singapore.

Storytelling

This over-imagined scenario of chaos and disorder is in many ways 'the Singapore Story'. As the State recounts it, the story tells of tumultuous struggle where brave and practical men – and less frequently women – fought against incredible odds and brought the modern nation into being. There will always be new challenges and new enemies, so there is no time for repose and complacency: each person – from

State leaders to people in the street – must be vigilant and act self-lessly to ensure the nation's survival and protect its prosperity. As the story counsels people on how to navigate through a dangerous and an unpredictable world, what are we to make of the storyteller? The answer is that he vanishes in his self-effacing conduct of sacrifice and devotion to duty. The first volume of Lee Kuan Yew's autobiography is called *The Singapore Story*.[5] One cannot read the book without being touched by a sense of the tragic. As Lee describes the fateful events that shaped his career and the Singapore nation, he comes across as a player in a drama filled with the larger scale of things, where everything was touch and go, and destiny almost took the political crown from him. To fate or destiny, even Lee is a servant.

One is here sharply reminded of German philosopher Walter Benjamin's insight in *The Storyteller*. 'In every case', he writes, 'the storyteller is a man who has counsel for his readers. . . . After all, a counsel is less an answer than a proposal concerning the continuation of a story that is unfolding.'[6] For the story to offer itself as 'counsel', however, the teller and the listeners must have sympathetic rapport. After all, the craft of the storyteller is to 'make [his experience] the experience of those who are listening to his tale'.[7] Storytelling thus makes a society of sort, connecting people by the life experiences the story evokes and by the enjoyments of telling and listening. In Benjamin's words, 'A man listening to a story is in the company of the storyteller; even a man reading one shares this companionship.'[8]

Much of this is going on in the telling of the Singapore Story. Of counsel, there is plenty in the two volumes of Lee's autobiography. Like Walter Benjamin, we have to think of Lee's sagacious words about commitment and sacrifice beyond their inherent political wisdom. What Lee proffers even more fervently is the irrepressible truth about the Singapore Story's continuing 'unfolding'. There is still more to be done after the work of national independence, after the big fight with communists and radical unions; Singapore's struggles and the PAP leaders' heroic endeavours are a tale that has no ending but always with another chapter to come. '[A story] resembles the seed of grains', to quote Benjamin's exquisite metaphor, 'which have lain for centuries in the chambers of the pyramids shut up air-tight and have retained their germinative power to this day.'[9] Told as a story, Singapore's struggle for survival lives on beyond the past where it first happened.

In this endless retelling, the story puts the State and the Singapore nation in cosy company and binds them to a common fate. And what they come to share is the crucial need to imagine the impending doom and to steer the nation away from it. The retelling of the story of hardship and sacrifice under a leadership of uncompromising moral vision: this is nation-building at its most elegiac. In the narrative of toil and blood, Singapore finds its highest expression in the great suffering of the past that still disturbs and pains. The logic of nation-building, and the need to harvest the moving effects would compel the retelling the Singapore's story in its final dramatic form, as something resembling trauma.

The idea of 'the traumatic', once we go beyond the modish psychologizing, best sums up the stressful urgency of the Singapore Story. History is the cornerstone of the story because it shows up the heroic endeavour of the PAP leaders in one form or another, giving them the moral capital they so ardently seek. When work pushed Lee to the point of a nervous collapse, as we shall see, it affirms the inner character of the man. The past is lesson for the present, just as imagining the worst a way of averting it. What the nation needs is a 'culture of excess', as I shall call it, that would ceaselessly alerts people to real and imaginary dangers and to find means to prevent them. 'Culture of excess' at the end comes to underline much of what the State does; it has proved to be of practical use even if it brings to the State's actions and decisions something like a distraught sensibility.

The 'culture of excess' has come out of the telling and retelling of the Singapore Story; as such it is a symptom of the dramatic, violent experience of national struggle – in the way people remember it. In its making as national history, the Singapore Story is experienced as trauma driving the unbearable need to retell the past sufferings and of course, their remedies. To rephrase Cathy Caruth's masterly meditation on memory and trauma, the State with its quick and excessive responses takes on 'the symptom' of a history that it has rewritten, a history once made real has a life of its own.[10] Before we come to that, we have to ask: What is this history that so traumatised the nation that it cannot help telling it again and again? Or to ask the question of Taussig: What is this history that 'emote[s] an abstraction', and renders the state as a 'thing' of 'passion and reason'?[11]

Taming the radical beast

After completing his law degree in Cambridge University in 1950, Lee Kuan Yew returned to Singapore to find a country teeming with nationalist aspirations and social unrest. After the Japanese occupation, British imperial power lost much of its lustre and real influence. There were stirrings of anti-colonialism and calls for political reform. The 'wind of change' sweeping colonial Asia, the Middle East and Africa, coupled with the weariness of war at home under a new Labour Government, drove Britain towards 'reformation and rationalization of the colonial presence in Malaya'.[12] With broad initiatives linking 'social economic policy with political intentions', the 'British reoccupation of Malaya was conceived as a vast experiment in democracy'.[13] There was growth of the public sphere, and lifting of the pre-war restrictions of voluntary associations, trade unions and the press.

The new political climate was also something of the political consciousness nurtured by the war. With the departure of the British, the Indian community had most famously lent support to the Japanese-sponsored Indian National Army (INA) to fight for India's independence. Radical Malays, on the other hand, looked to Indonesia for inspiration, where at the end of the war Sukarno's guerrillas were fighting against the Dutch. For the Chinese population the Sino-Japanese war, which started with Japanese landing in Tientsin in 1938, had been the crucial political training ground. As the war spread to central and southern China – the Japanese army took Canton the same year – the news of devastations and massacre of civilians prompted Malayan Chinese to organize boycotts of Japanese goods, and relief projects that collected funds and materials to be sent back to the Mainland. In these activities, local organizers 'jostled with Kuomintang and Communist elements'.[14] The war efforts energized a new Chinese patriotism, and more importantly, a political sensibility built on the experience of mass mobilisation and modern, nationalist feelings.

The Communist-based organisations had a special place in post-war Malaya and Singapore. Enjoying considerable popular appeal, the Communists formed the only effective guerrilla force against the Japanese during the dark years of 1942–45. The retreating British had sought collaboration with the Malayan Peoples' Anti-Japanese Army (MPAJA), the military wing of the Malayan Communist Party (MCP), in organizing a stay-behind party for gathering intelligence for the Southeast Asia Command in Ceylon. On its part, the MCP had hoped to strike a bargain with the British and perhaps to carve out a legitimate political role in post-war Malaya. In any case, in the relatively liberal climate after the war, organizations with varying degrees of Communist influence grew; so did labour unions, student societies and trade associations whose diverse ideological aims could be loosely described as anti-colonial, multiethnic and democratic-socialist. It is problematic to label these organizations as communist, so is tracing their direct or tacit links with the MCP and its off-shoot the Ex-Comrades Association. The historian T. N. Harper is most accurate when he describes these associations as underpinned by 'a common ground between the "Eight Points" of the Malayan Communist Party, the republicanism of the Malay left, and the global solidarities of revived trade unionism.'[15]

So it was this heady atmosphere of anti-colonial nationalism and socialist radicalism in which the young Lee found himself. But Lee's political ideology was of a different hue. At London University and later Cambridge, his major influence was moderate, Labour Party 'reformism and social reconstruction' which came to provide the founding vision of the PAP's struggle for independence.[16] Nonetheless Lee was deeply impressed with the energy and commitment of the students and labour organizers with whom he came in contact. Politicized by the war, the militant men and women presented a dynamic political force that Lee and his colleagues could not be ignored; here is his passionate assessment:

> ...one day in 1954 we came into contact with the Chinese educated world. The Chinese middle school students were in revolt against national service and they were beaten down. Riots took place, charges where preferred in court... We bridged the gap to the Chinese educated world – a world teeming with vitality, dynamism and revolution, a world in which the communists had been working for over the last thirty year with considerable success.
>
> We the English educated revolutionary went in trying to tap this oil field of political resources, and soon found our pipelines crossing those of the Communist Party... We were considered by Communists as poaching on their exclusive territory.[17]

The commitment and dedication of the left was the more impressive when brought in contrast with the English-educated who like Lee himself benefited most from colonial patronage. In the war years, against the great suffering of the Chinese-educated who bravely put themselves behind the wheel of resistance, the Westernized elite was distinguished by ingratiating themselves to the Japanese, so Lee remembers:

> Many of us will remember the unhappy spectacle of English-speaking, Western-educated colleagues suddenly changing in their manners of speech, dress and behaviour, making blatant attempts at being good imitation Japs. Indeed some were sent to Japan so as to be better educated, to enlighten their ignorant countrymen in

Malaya and doubtless also to become the privileged class, second only to the genuine Japanese themselves.[18]

For Lee the greatest sins of the English-educated lie in their self-interest, and failure to cast their lot with the anti-colonial movement. He was certain that Singapore's political future would be in the hands of the Chinese radical left with whom moderates like him and his colleagues had to form a political alliance.

The Hock Lee riots

The PAP was inaugurated on 21 November 1954 by Lee and other 13 leaders in a meeting attended by 1,500 members. From the on-set, the PAP took a strong anti-colonial stance, demanding national independence through constitutional means. It later called for merger with Malaya with the view of creating a national movement by bring-ing together all anti-colonial forces on both sides of the Causeway. Slowly Lee and his colleagues began building a party with a mass base. For this they relied on the trade unions and Chinese school activists who brought working class supporters. The period 1954 to 1961 generally saw the collaboration between the PAP moderates and the radical unionists. But the relationship was uneasy and often volatile. It reached a climax in June 1961 when the radical elements led by the charismatic unionist Lim Ching Siong broke with the party and called for the PAP government to resign; Lim together with radicals were later arrested in a security operation in September 1963.[19]

The history of Singapore from 1954 to 1961, a period covering the founding of the PAP to the departure of its leftwing faction, is generally described by the PAP circles as one of 'astride [the back] of the tiger.'[20] A popular cartoon book depicts Lee with his legs across the ferocious beast, the right fist in the air clasping the PAP party emblem, ready to punch a blow on the head.[21] His clothes in tatters, face grimed with determination, it is the visage of a man in the midst of battle as he strikes terror against the enemies visibly cowering under the tiger's belly. Melodramatic at best, the cartoon celebrates the heroism and danger of the PAP enterprise. When we recall Lee's deft machination in harnessing the leftwing unions to ensure the PAP's electoral success, 'crushing the tiger' also speaks darkly of betrayal and parasitic undertaking of Machiavellian genius. In any case the narrative cannot but turn to the other side, telling of PAP's political vulnerability, as much as the dangerous undertaking of taming the radical beast. It was no 'paper tiger' on whose back Lee and the PAP rode to power.

Lee's 'praise and blame' ambivalence of the radical left is shown up in his first meeting with Lim Chin Siong in the Chief Minister David Marshall's office. Pointing at Lim, he had said to Marshall, 'Meet the future Prime Minister of Singapore!' 'Don't laugh! He is the finest Chinese orator in Singapore and he *will* be our next prime minister!' he added.[22] In a moment of unguarded mirth, Lee gave a sharp evaluation of his nemesis. The political resourcefulness and commitment of the world that Lim inhabited were both dangerous and something that PAP desperately wanted. Enjoying popular support unionists like Lim were able to call for mass strikes and, as was often accused of them, violent confrontation with the police. In the PAP narrative, the students and workers barricading themselves in the schoolyards and the factories were the 'radical beast' out to destroy the world it tried to bring about. Among the narration of blood and loss of life, one event has achieved almost iconic status – the 1955 Hock Lee Bus Depot strike.

In Singapore 'the Hock Lee riots' – to use the official phrase – call up scenarios of industrial standstill, mob violence and unruly labour unions. It was a time when the PAP was struggling to consolidate power and prepare for the country's self-rule. Indelible images of the riots are reproduced in numerous books and magazines, and at the Singapore History Museum. There visitors find disquieting images of bus workers with raised fists voting for the strike; Chinese pupils in white shorts and skirts addressing the crowd; policemen training a powerful fire hose at the picket line outside the bus dept; and most evocative of all, the street parade of the funeral of Chong Loon Chuan, a young student injured in the riots who later died from the wounds. In the official view, the strike had been inspired by communists in order to cause violence and unrest, and discredit the British authorities and the anti-communist local government led by David Marshall. At the time Lee was the legal adviser of the Singapore Bus Workers' Union, and Lim Chin Siong and his associates were still a major faction of the PAP. This is how old Asia-hand John Drysdale describes the strike:

> At dawn on the morning of April 21 the police attempted to break up ... pickets outside the bus depot. Fifteen of the strikers were injured. By 9.30 a.m. over 1,000 workers and students gathered outside the bus depot and decided to picket the gates before dawn on the following morning. . . .

On May 8 the police riot squad was brought in to disperse a crowd picketing the depot. Some buses were running but they were being boarded by strike sympathisers who ripped up seats and persistently rang the bells. Two days later, police, disguised in plain clothes, were posted around the depot, and hoses were used to disperse the pickets, nineteen of whom were injured. An emergency meeting of the Factory and Shop Workers' Union was called by Lim Chin Siong to protest against the use of force and to declare a general strike if the Hock Lee dispute was not settled 'reasonably'. The strikers, encouraged by 800 Chinese Middle School students, issued a statement that they would prefer imprisonment if the management did not reinstate them.[23]

By 11 May the Bus Workers' Union gave an ultimatum to the Hock Lee management for it to settle the dispute in 48 hours. The police began to close in.

Events in the early morning of May 12 confirmed Lee's worst fears. The pickets manning the gates of the Hock Lee bus depot shouted defiance with clenched fists as the police closed in on them. Fong Swee Suan [one of the union leaders] urged the pickets, if they were 'brave enough', to stand firm, but the fierce jets of water from fire hoses swept the pickets away. The buses passed through the gates pelted by stones thrown by students and others who had gathered in their hundreds along Alexandra Road. By the afternoon, students of the Chung Cheng High School and the Chinese High School converged on Alexandra Road circus in 20 lorries. A pitched battle between police and 2,000 rioting students, 300–400 strikers and others, raged from the Hock Lee depot to Tiong Bahru. Stones, bottles and tear-gas were the main weapons of the day though the police also used firearms.[24]

Running battle continued until the next morning, and some 60 Gurkha police were called in to disperse the crowd. When security forces finally crushed the rioters at dawn, four people had been killed and 31 injured.

For the PAP the Hock Lee riots revealed the true face of the unionists, and gave its leaders their first experience of 'the potency and uncontrollability of the Communists in [the] party'.[25] Lee himself was pushed into a corner. Aware of the violence that the radical left could unleash, envious of the mass support it enjoyed, Lee nevertheless had to defend the unionists for the sake of party unity. Besides, openly condemning his leftist colleagues would be seen as throwing weight behind the British and colonialism, something that would undermine PAP's popular appeal. Lee, the reluctant suitor of communists, made his position clear in the Legislative Assembly:

> The Chief Secretary wants to know what I would say, without quibble, about communism and colonialism....I say that I will not fight Communists to support colonialism.
> Not all the riots will make me do it. But I say here and now that if I had the choice between democracy – an independent, democratic Malaya, a Communist Malaya, and a colonial Malaya, I have no hesitation in choosing and in fighting for an independent, democratic Malaya.... We will not fight the Communists or other Fascists to preserve the colonial system.[26]

The ideological bind clearly unsettled Lee. The challenge facing the PAP was tussling for mass political base so that it could be the inheritor of power when the day of independence came. This called for, for the moment, ways to accommodate the radicals, even defending them against the British. The labour unions were the thorn in the PAP's political crown, but also a reservoir of political support it could not do without.

That not how the State narrative would tell it, however. Rather the labour unions – and events like the Hock Lee strike – should serve to remind people of the violence, disorder and loss of life in Singapore before the PAP came to power. And in this narrative, the Hock Lee strike adds to other critical happenings in the early years of national struggle. As school children have come to learn them by heart, these are: the economic disaster following the British withdrawal of the military bases; the race riots between Chinese and Malays; Indonesian President Sukarno's *Konfrontasi* campaign to topple the newly formed Federation of Malaysia; the Chinese students' demonstration over conscription and other issues; and of course, the heart-wrenching disappointment of Singapore's expulsion from the Federation. These events make up the nation's tumultuous past, the spell-binding proceedings of its birth.

The Hock Lee riots have a special place in the Singapore Story. As a part of the PAP faction, and before their repulsion in July 1961, union leaders like Lim Chin Siong and others were a kind of 'enemy in our midst'. The PAP both courted and feared them. For Lee, the Western-educated elite too prone to kowtow to the British were pathetically 'irrelevant' in the anti-colonial struggle; labour unions and the Chinese-educated world were something else altogether. The political genius of Lee had been to realize this, and went on to harness the political resources and electoral support of the left, and did away with them once the PAP had built its power. History has shown how brilliantly Lee been accomplished this. It is no less an accomplishment in installing the bloody Hock Lee riots deep in the popular imagination. For Singaporeans, the event not only signifies the unbearable fate of Singapore without the PAP, but also poses a warning. For the radical left and labour unions that brought havoc in those anarchic days were to follow by other malevolent figures. In assessing these figures – unionists, communists, ethnic chauvinists, religious extremists and Catholic NGO's seduced by liberation theology – there can be no repose and calm judgement. It was all blood and anarchy, impetuous youth and misguided poor instigated by communists out to seize the world by violence. It is this memory, constructed as something akin to trauma, that has given rise to the 'culture of excess' in Singapore.

History as trauma

The idea of 'trauma' reveals important things about the Singapore Story and the way it is being told and remembered. It shines a light on the psychological tenor of the State actions: its moralistic bend and anxious posturing, its heightened imagining of social and economic doom, and its 'over-responses' to crises. Dramatic and heart-wrenching, the narrative is also banal and repetitive, sentimental and arousing; features that give the recollection of history a 'surprising impact' yet making it out of reach as experience. All these are captured by the notion of the 'traumatic'.

On the subject, Cathy Caruth has vividly written:

> ... most descriptions [of the traumatic] generally agree that there is a response, sometimes delayed, to an overwhelming event or events, which takes the form of repeated, intrusive hallucinations, dreams, thoughts or behaviors stemming from the event, along with numbing that may have begun during or after the experience, and possibly also increased arousal to (and avoidance of) stimuli recalling the event.
>
> This simple definition belies a very peculiar fact: the pathology cannot be defined either by the event itself – which may or may not be catastrophic, and may not traumatize everyone equally – nor can it be defined in terms of a *distortion* of the event, achieving its haunting power as a result of distorting personal significances attached to it.[27]

The traumatized person is compelled to 'remember' the catastrophic event, and yet he must avoid the emotional impact of it. He is possessed by an image or event, and the need to revisit it. Against his will, the traumatized person returns endlessly to the site of the originary experience. In doing so the person must confront the 'truthfulness' of the event, yet resist completely knowing it. In this sense, trauma is experienced, not only as distortion, but also as something affirmed by the senses here and now – months or years or decades after the events had occurred, at another place. Despite the time and distance, the experience is for the traumatized pressingly real. What floods into her present recollections is imbued with '*literality* and [a] nonsymbolic nature', in Cathy Carruth's word.[28] The traumatized cannot help going back to the past any more than he can truly 'live' in the event that so distresses him.

In short the idea of trauma is about remembering, the uncertain yet predictable way in which a painful event is recalled and relived. It is above all about history and its distortion, and how people are made the suffering symptoms of it. In these terms, the idea begins to reveal something of the deep entanglement of the Singapore Story.

If an event like the Hock Lee riots has a compelling relevance today, it is precisely because it has been artfully retold as the trauma of Singapore's birth as a nation. The Hock Lee riots may be a distortion, they are nonetheless remembered – and experienced years on – as real. Here we begin to see the social effects when Singapore's national history is rendered as trauma. For to the traumatized subject, history is both symbolic and real. In Cathy Caruth words, 'To be traumatized is precisely to be possessed by an image or event', and memory is 'purely and inexplicably, the literal return of (the event or image) against the will of the one it inhabits'.[29] The Singapore Story is filled with the same driving compulsion. To the national subject, the Story always comes across as something of a 'historical enigma',[30] packed with real events and psychologically exacting experiences, yet evident of the storyteller skilful distortion. An event like the Hock Lee riots make sure that history is experienced as real, yet charges it with the frightening qualities of nightmare. A festering wound in memory, it incessantly reminds Singaporeans of past dangers and bloody chaos they cannot afford to live through again. As the State reworks the past events into the most deeply felt, it infects them with the 'crisis of truth' and in the process people are made 'symptoms' of their memories.[31]

If national memory is replete with anxiety, from the view of nation-building this is not altogether a bad thing. For the State, retelling the Singapore Story hopes to imbue the national subject with this anxiety, who is compelled to endlessly revisit the site of blood and violence, and to imagine the nation's collapse as an ever-present likelihood. No wonder 'taming the radical beast' has to be staged with such self-possessive drama. If the Singapore Story produces a 'traumatic experience,' the State is likely to see it more desirably as generating a nervous energy that prepares the nation for disaster. 'Over-responses' make people forever sit on the edge of the seat, and this, we might say, is exactly what the State wishes. Worrying to excess – restlessly pondering the uncertain future – is the balm for the 'traumatic mind'; it is also, in another realm, a collective virtue that Lee personally exemplifies. In this sense, anxiety is more than what it is but a mainspring for action and endeavour.

Democracy: 'the empty space of power'

The idea of 'the traumatic' explains a good deal of the stressful qualities of the IHT-Lingle case and other events that will unfold in the book. The IHT-Lingle case – its reasoning and the State's quick and rigorous response – is both a symptom of a 'traumatic experience' and a defence against it. And we can speak of the State's frequently 'over-the-top' responses in this and other cases as forming a 'culture of excess'. As it infects the State and the everyday life, the 'culture of excess' is at once debilitating and useful. Like the experience of the 'traumatic', the 'culture of excess' traps people in a history that eludes them; but it also 'incites' them to prompt actions as a way of avoiding the similar, often imaginary disasters. For the State, to be 'excessive' in an event of crisis is pertinent and practically necessary. Such a gesture strings the current crisis with past events, and spurs the State to quick action.

This extravagance of action and responses has become something of a national ethos; and much of this book is given to showing the unmistakable, feverish manner in which the national ethos reveals itself and shapes people's life. For the State, the 'culture of excess' is firmly settled on the pragmatic side, emphasizing as it does efficient administration and public duties. And it is through the quick actions and nervous energy that the State renovates itself. As it returns to the traumatic events of the past, as it anxiously pursues the recalcitrant Other whose unruly ways signal what Singapore should avoid, the State also turns its scrutiny to another field, one even more fertile for the seed of social chaos.

For its admirers and detractors alike, Singapore is famous for its 'fine tuning' of the system of parliamentary government inherited from the British. Ever since independence, the PAP government has never held back its view on the inadequacy of Western liberal democracy. The 'socialism that works' of the 1960s and the 'supply-side socialism' in the later usage have come out of the practicalities of political rule and PAP's ideological ambitions. In modifying many of the features of liberal democracy – from the right to strike to the role of the political opposition, individual liberty to communal responsibility, human rights to the free press – the Singapore State has always insisted that it has done so in order to adapt an alien form of government to the Asian condition of different needs and cultural traditions. Much of this has to do with the State's need to ensure its continuing hold to power; of that there can be no doubt. Nevertheless, the State's fine-tuning of liberal democracy is more suggestive than that. Generally, the attempt also tells of the uncertainty of postcolonial states in their transition to nationhood.

Quite apart from rebuilding the economy, new nations faced some stark realities: the change of political status quo and the loss of bureaucratic machinery and technical services provided by colonialism; in some cases, as in India and Angola, colonial regimes would depart in haste and leave the newly found nations in social and political chaos or ruin. The great social upheaval of national struggle, even as it elicited fervent political hope, was by the very nature 'traumatic'. Speaking of postcolonial Asia and Africa, the chaotic experience of decolonization may well explain why many nation-states had turned to the 'totalitarian form' of various shades. In post-British Singapore, we see the 'hollowing out' of its cultural and political centres. The building a 'socialism that works' is about regaining the pivoting underpinning we associate with social-economic stability and national order. The anti-colonial struggle may be the stirring stuff that warmed a revolutionary's heart; new independence also lit up the many practical tasks that awaited the new regime. In Singapore the withdrawal of colonialism was to leave a gaping hole in the economy, and in military security which Britain had been providing. The PAP leaders were not given to festive euphoria for their new victory and quick to remind the people of the difficulties and dangers ahead: when it comes to that, national independence was a mixed blessing.

We have encountered Lee Kuan Yew at the point of nervous collapse from the strenuous tasks of the new government. The national father fallen sick is a powerful metaphor of the excruciating uncertainty in the interim between the old and the new regime. In European political philosophy, the most famous story of the national father in demise is the French Revolution. That narrative of revolution turned bloody tyranny, in spite of its special history and circumstances, helps to illuminate the tension and moral calamity facing a postcolonial state like Singapore. Our intellectual guide is French political theorist Claude Lefort, particularly his two essays *The Logic of Totalitarianism* and *Image of the Body and Totalitarianism*.[32] In the *ancien régime*, Lefort tells us, the king's body was closely identified with France and stood for its secular and spiritual authority. The imagery, drawn from medieval idea of kingship, tied the king's body to that of Christ, so that the 'secular king' became, like the Son of God, both mortal and eternal. The royal body ultimately came to symbolize the perfect coupling of sacred kingly power and secular national community.

This coupling was ruptured by the Revolution of 1799. The execution of Louis XVI spelled the end of the 'family romance' in France which installed the king as the powerful but caring Father.[33] The effects of regicide were both political and psychological. As the guillotine sliced off the head before the revengeful crowd, it let loose the nation from the Father, and thus from the communal and existential anchor of society. And this is Lefort's insight: if regicide was necessary in realizing the ideals of the Revolution, it also left the centre of political power as an 'empty place'.

The tragic blood letting of the French Revolution is for Lefort a powerful sign of democracy's inherent instability. As a political form, modern democracy is generally 'characterized by the experience of society as "historical," in constant flux – and as marked by social division, not homogeneity.'[34] In the conditions of flux, the need to re-fill the 'empty space of power', to realign the national body with the community, becomes a significant tension. Lefort writes:

> [When we] speak of the disincorporation of the individual . . . we must examine the disentangling of the spheres of power, law and knowledge that takes place when the identity of the body politic disappear. The modern democratic revolution is best recognized in this mutation: there is no power linked to a body. Power appears as an empty place and those who exercise it as mere mortals who occupy it only temporarily or who could install themselves in it only by force or cunning.

In a democratic regime, freedom is experienced as a kind of horror:

> There is no law that cannot be fixed, whose articles cannot be contested, whose foundations are not susceptible of being called into question. Lastly, there is no representation of a centre

and of the contours of society: unity cannot now efface social division. Democracy inaugurates the experience of an ungraspable, uncontrollable society in which the people will be said to be sovereign, of course, but whose identity will constantly be open to question, whose identity will remain latent.[35]

The weakness of democracy lies not only in its structural instability but also in its spiritual void.

Totalitarian ambition

When power is seized from the oppressors, and in the new dawn of hope and equality: to whom does this power belong? The romantic vision of revolution would have us answering: it belongs to the people, to everyone. In democracy power is not longer immortal but finite, and it has to be periodically renewed through the popular mandate in order to incorporate 'people's will'. For Lefort the tragedy of democracy rests on its particular form of power: it belongs to everyone and yet to no one. Democracy is not so much 'an empty place *of* power, [but] an empty place which *is* power.'[36] The empty place is not the absence of power, so much as the site of the anonymous 'everyone' who is the rightful owner of that power. But the mythical 'everyone' is really 'no one', and with this, the culture of democracy, emphasizing as it does the fragmentation of power across civil societies and the public sphere, looks increasingly like an invitation to chaos.

And this is Lefort's great theme: democracy – that experience of an 'ungraspable, uncontrollable society' – tends to carry the seed of totalitarianism. For many, the attraction of totalitarianism is precisely its promise to overcome the unbearable 'indeterminacy' of 'power for everyone', to refill the existential void by devotion to the Leader. If democracy is burdened with the contingency of history, the totalitarian system is 'a society without history' just as it creates a 'new man' that is mythically pure and eternal.[37] The totalitarian system is not easy to bring about, requiring as it does the right social and political conditions, and not the least, charismatic leadership and organized means of violence. But the desire for it, and the ambition to bring about some of its features, is always enticing. In many postcolonial regimes in Asia and Africa, when we cannot speak of the totalitarian form of a Mussolini or a Stalin, we can nonetheless point to their evident, barely disguised ambition to achieve some elements of it. Such ambition – and the practices and policies it drives – are seductive panacea for the shortages of democracy. For the postcolonial state like Singapore, traumatized by history of its own telling, it is only too tempting to turn to a political form in which conflict and division have no place, where national history is also the history of the ruling party under a leader of powerful personality.

Lefort's meditation on European medieval kingship and the totalitarian regime is light years away from tropical Asia, yet it resonates with the problems that obsess the Singapore State. The island republic, like neighbouring Malaysia, has a Westminster parliamentary system of government. However many practices of its practice would be hard put to measure up against normal liberal-democratic standards. These practices, as it has been argued *ad nauseam*, are driven by the pragmatics of efficient government, and the need to modify an alien, European political form for Asian conditions. In a way, one cannot begrudge the argument. A postcolonial regime like Singapore had to start anew, and not be burdened by the old forms and practices. Some like compulsory acquisition of land for public housing at state-determined prices, and notably, and control of vehicle traffic by a huge tax levy on car ownership are clearly of benefit to the society.

Political and administrative practicalities however do not exhaust the significance of the State's actions. Nation-building in Singapore often seems like the manufacturing of panacea against the imagination of impending doom. But it is also true that regional instability and political chaos created by radicals and religious fanatics alike are sometimes pressingly real. Communist insurgency, and more currently September 11 and the Bali bombing had proved that the State might have been right. Nonetheless when the national narrative is full of scenarios of doom, when perpetual crisis the central feature of the dark, anxious 'culture of excess', then what is real is already something of the over-wrought imagination. This mixture of invention and the real, one may say, gives the national narrative both social appeal and psychological urgency. The Singapore Story turns out to be a beast of uncertain will. It forewarns, and acts as a kind of pre-emptive strike, but it cannot do this without spicing up of all kind of details. The Singapore Story as trauma is merely the most colourful of the discursive invention. And what is the 'traumatic' except the experience that frightens and ensnares a person into an 'impossible history', even as it spurs him into action.

Viewed this way, the 'culture of excess' really suggests the State's need, not so much for totalitarianism as such, as for the ambition to achieve and utilize some of its formal features. For Lefort lively civil society and plural political visions are important ways of overcoming the terror of democracy's 'emptiness'. But he has no delusion

that totalitarian measures are a tempting shortcut for regimes impatient with democracy's clumsy limitations. And these limitations
are, for Singapore, both moral and political. The State's ideological
projects – from the 'socialism that works' to Asian Values and Shared
Values – are attempts to create for the nation an indelible existential
centre. Invested in the 'community' and 'social consensus', the highest
ideal of the nation always insinuates a fantasy that all 'criteria of law
and of knowledge' can be brought within of a single, unitary mode
of power. Asian and Confucian Values and 'democracy with an Asian
face' are very much that. And indeed, what the State would like to
have for Singapore – some put in place, some too impracticable and
later abandoned – seem remarkably like what Lefort describes as the
'totalitarian form': 'the model of a society which seems to institute
itself without divisions, which seems to have mastery of its own organization, a society in which each part seems to be related to every other
and imbued by one and the same project of building socialism.'[38] This,
of course, is not to associate the PAP State with that of Mussolini
or Stalin; 'ambition' and 'aspiration' are suitable, moderate words
to describe the measures of the Singapore State. Cautious of unfair
comparison, we should recall that over the past decades, PAP leaders
have never shied away from expressing what they see as the structural
and spiritual malaise of liberal democracy, and their desire to find
a remedy. The fine-tuning of the parliamentary democratic practices
is very much a fact of political life in Singapore. This can even be
a gesture of postcolonial pride, a slap in the face for the tired, old
regimes of the West. As for the elements of 'totalitarian form' that
Lefort outlines, there is a great deal in them that the PAP State would
admittedly find attractive and useful – though it would use a different
language to describe them. The ideas of the state and society as one,
of national community above the individual, and collective goal above
private desire are stuff of nation-building, so it is claimed. They are not
meant for creating a system of the former Soviet Union, but to make
democracy work better. At the end, there is no way of associating Singapore with the unsavoury regime; economic prosperity and efficient,
honest government undercut in no small way the ethical problems of
the State's totalitarian ambitions.

History and the 'culture of excess'

In Ridley Scott's *Black Hawk Down* about American military mis-
adventure in Mogadishu, the US commander Garrison has 'invited' a
powerful local man to his headquarters. Over Cuban Bolivar Belicoso

cigars and 7Up, Garrison asks the whereabouts of the warlord Aidia and conversation leads to the American mercy mission in Somalia. With the sweep of his hand, gesturing toward the famine and lawlessness outside the comfort of the army base, the visitor taunts Garrison: 'Don't think because I grew up without running water I am simple.' 'I do know something about history'; he then digs the knife in, 'See all this? It is simply shaping for tomorrow. A tomorrow without a lot of Arkansas White Boy's ideas in it.'

History, as it unfolds, is ours; it belongs to its makers. For all its violence, the history of starvation and warlordism is still a thing of the Somalian people, and no amount of America's force-feeding of democracy can change that. They would ride on its wings, so to speak, and follow where it takes them even if the journey is to end in genocidal horror. This is simply the Somalian way guided by its own impeccable logic. To speak of 'our history', however, is already suggesting that it is something of our doing. The making of history is of course the great romance of all revolutions. And this romance, when we think of it, is not only one of changing the old regime, but also of telling another story about *our* lives and *our* aspirations. In Somalia as elsewhere, history is the stuff of the narrative of the present and future.

In Singapore, the telling of the nation's story is similarly compulsive. It makes a great deal of PAP's good work and helps to map out its ideological vision. It dramatizes the chaos and struggles of the past in order to energize the present endeavours. And not the least, it becomes the knowledge of the everyday, defining the horizon of hope and moulds people's responses to the State. When past events are made into a lesson, when they are retold to emphasize their painful experiences, the result can be bleakly paradoxical. For Singapore, the nation's story as trauma is merely collective struggle told in heart-wrenching and awe-inspiring ways. For visitors too, the island republic enjoys some striking understandings, not all flattering. She is at once the only First World country in Asia after Japan and a soulless nation of materialist pursuit; to science-fiction writer William Gibson, Singapore is 'Disney with death penalty'. Behind all this and the shopping malls and smooth traffic, and tough laws and efficient government, lies an anxious national ethos that makes Singapore tick.

The 'traumatized' national ethos may suggest a warped pathology; but since it makes people work hard and has transformed the island into what it is today, not a few Singaporeans are proud of it. The 'culture of excess' when shorn of its fretful qualities is simply a counter-measure against social indolence and moral lethargy. For the more reflective outsiders, one of the most perplexing aspects of Singapore – besides the restriction of chewing gum sales and until recently, banning of bar-top dancing and bungee jumping – is its political form which none of the terms like democracy, authoritarianism and totalitarianism can adequately describe. Singapore is perhaps all these things. Critics of the Singapore State have to reckon with the fact of the island's economic success and the PAP electoral victories term after term, however complex the reasons for them. The anomaly cannot be easily swept away for the sake of a progressive critique. What happens on the island is not all death of democracy and ideological mystification; popular support if not complicity is a feature of Singapore's political life too.

The 'totalitarian ambition' of the Singapore State is similarly entangled. I am not sure if the brilliant men and women in the inner PAP circles have read Lefort and other writings of Post-Heideggerian thoughts. But Asian Values and Confucian Capitalism and the creative adjustment of the Western liberalism are seemingly remarkable attempts to rewrite democracy and its limitations, and to remedy the moral void. This too is the magic of the Singapore State: its effort to fill in the 'empty place' of democracy with the wholesomeness of the community, with the ethical force of post-Cold War New Asia.

If 'totalitarian ambition' is the way for rewriting Singapore's own history, it is also something with which we use to evaluate its achievements and moral credential. Struggling to 'fix' society from historical flux, political aspiration of this sort is always marked by the tendency to deny 'that division, conflict, and antagonism are constitutive of society'.[39] Historical flux, and 'division, conflict, and antagonism' are also the nemesis of the Singapore State. The tragedy of Singapore, we might say, is precisely the imposition by the State – not without some popular support – of a unitary vision of a good society, and the denigration of an alternative political future beyond what is normally imagined. What lie outside the PAP State are realities of chaos and social and economic collapse. Security is bought with ideological and moral short-sightedness. This is not to suggest that the PAP vision of things is universally static, or that it has remained the same now in the hands of younger generation of leaders. With minor adjustments made necessary by the new and the contingent, the fundamental PAP vision will remain to guide Singapore: the 'culture of excess' will persist to posit a fixed, originary point to which the anxious national ethos can endless return.

And we do not need the language of psychology to see what is happening in Singapore. Certainly, the PAP government would put a much more positive spin on the ideas of 'excess' and 'the traumatic' even if these are not the words they would use. Still once removed of their pathological nuances, these are simply other terms for dynamism and inventiveness. In economic affairs, they speak of skill upgrading, searching new markets and creating value-added IT (information technology) to bring the economy to the twenty-first century. Now more than ever, creative thinking and innovation are the call of the day, announcing as it does the new imperative for the globalized economy.

However, creative thinking and calls for political openness always sound dangerously like 'division, conflict, and antagonism'. Hence we witness some strange double-moves today: gay men can now be employed in the civil service, but the law has not been changed to make homosexuality legal; the chewing gum ban has been lifted but you can only buy it from a pharmacist; in the local 'Hyde Park corner' speakers need to get a police permit to address the pathetically small crowd; and so on. The building of the new opera house has to justify its huge expense with the reason of making Singapore a more civilized city to attract foreign executives to come and work here; so pragmatism still apparently rules the current softer State rule. Openness, Singaporeans are reminded, has a cost. Creativity is not inherently for itself, for the excitement and intellectual enrichment of it. Practical reasons are still important, and the alternative is wastefulness or disorder. The effect of the continuing hold of PAP rule is always embedded in a traumatic vision and anxious national ethos. For all the talk of values and ethics in the State's scheme of things, it is an approach lacking in moral insight – whatever the eventual economic outcome.

3 'Yellow culture', white peril

Within the dense welter of our material life, with all its amorphous flux, certain objects stand out in a sort of perfection dimly akin to reason, and those are known as the beautiful.

Terry Eagleton, *The Ideology of the Aesthetic*

Two bodies

The pose is languid and a touch defiant. They lean against the wall with their arms barely touching, behind them the Cambridge University landmark the Bridge of Sighs. Lee Kuan Yew and his then girlfriend Kwa Geok Choo had been law students at the University since 1947; they were young and in love. Their quiet repose exudes a delicate intimacy, and a certain social bearing nurtured by family background and the best of British tertiary education. Here in Cambridge, before Lee's graduation and return to Singapore in 1950, there was still time to relax and enjoy what any young man about town of reasonable means might fancy. The picture dated itself to 1948. War-worn London might not be the centre of chic and fashion, but it nonetheless held the Olympic Games that summer, and the 'American Broadway Invasion' brought *Oklahoma!* and *A Streetcar Named Desire* to the theatres. Of these pleasures Lee makes no mention, preoccupied as he was with his final examinations. Nonetheless, this and other pictures included in his autobiography *The Singapore Story*[1] do show something more than his usual, stern demeanour. The slick hair and the sartorial elegance of tie and woollen waistcoat under a jacket – a picture taken outdoors in the winter of 1947 showed him with a cigarette in his right hand – would have suggested, if not for the academic gown and the Cambridge backdrop, something of a dandy. There is in these postures a quiet, youthful defiance, the

romantic indolence of a James Dean or a Marlon Brando or a character from the deadly fashionable films of the Hong Kong auteur
Wong Kar Wai.

Some ten years later in 1959, on a Friday afternoon in June,
Lee and his team walked to the City Hall to be sworn in as the
first self-government of Singapore after winning a 43 out of 51
seats electoral victory. All were dressed in white, the men in open-
necked cotton shirts and trousers, the few women in Chinese blouses
and long trousers. It was recognizably the attire of the Chinese-
educated world whose dynamism and mass support had helped the
PAP rise to power. Students, teachers and labour activists had taken
to wearing clothes of white starched cotton not only for their cool,
airy practicality in the tropics, but also for their 'ideological fashion
statement'. Pregnant with signs of austerity and sombreness, white
cotton was indeed *the* attire of the politically committed for whom
dark Dacron, not to mention Elvis Presley and Cliff Richard, was
Western decadence or 'yellow culture' reincarnate. Of this the PAP
leaders were all too keenly aware. Apart from official functions, they
would take to dressing in immaculate white even in a tree-planting
ceremony or when they took part in a street-cleaning campaign. These
were 'copycat exercise[s] borrowed from the communists', Lee states,
exercises that mimicked the energy and commitment of the radical
left. Mobilizing 'everyone including the ministers', such public works
signified 'service to the people' by the toiling with hands and soiling
of clothes.[2]

For PAP leaders like Lee, there were plenty of reasons for these
exercises of labour and discipline. To gain the mass support of the
labour movement was one; but as they looked to the West, they found
visible evidence of social decay that served as warning for Singapore.
Take for example, the scene of *la dolce vita* that Lee had witnessed
during his visit to Italy:

When I was in Italy in 1957, everybody – it was the age of
the scooter – everyone had a scooter. Five years ago, all Vespas
running around. This time I went there and the first thing I
noted was all the scooters had been replaced by little Fiats,
600, 500, and chaps who've got Fiats don't go and embark on
revolution. They are thinking of the next instalment, how to
make sure that they've got the next instalment to pay the Fiat
dealer.

On Sunday, Lee and his party took time off from official duties for an outing to the countryside:

> We went out to the country on Sunday...there must be 100,000 families with the same ideas. They also went out, everybody with a little Fiat or an Alfa Romeo.... And everybody brought a little tent or a fishing rod...if they were young they made love, if they were old they just sat down under the sun and sipped mineral water.[3]

As usual, Lee would inject even in this scene of popular idyll something of his didactic vision. 'Chaps who've got Fiats don't go and embark on revolution' sounds like a slip of the tongue; but it suggests, in Lee's lively imagination, a lesson of political and moral import. The possible lesson for Singapore is surely that which he saw in abundance in Italy: when people were satiated with material goods and social enjoyment, there would be no place in their lives for passionate idealism or irrational political demands. People with jobs and a full stomach, to rephrase Lee, do not throw barricades in the streets.

But there are costs to all this. As social peace and economic prosperity begin to do away with political slogans and mob politics, they tend to be taken as a 'right' to be enjoyed after the toil and turmoil of nation-building, Lee would perhaps imply. Viewed in this light, an Italy too fond of fast cars and love-making and lying on the beach cannot but pose a lesson for Singapore. As it matures and transforms into a First World society, the task is to prevent the development of soft, feeble bodies that value relaxed indolence above all else and refuse to take the economic as the only measure of things. For Singapore the solution seems clear: it must nurture among the people a national culture of discipline, tough-mindedness and pragmatic calculations, one deeply ingrained with the ethos of bodily discipline and moral magnitude.

In its short history, modern Singapore has been obsessed with fostering such a national culture. As in any project of this kind, it celebrates the collective struggle of the people against enemies of one form or the other, and the wisdom and heroic endeavours of the political leaders. History, suitably retold and dramatized, aids the imagining of the future. And this imagined future is always presented as a plan, a template or, better still, a map with which the nation can navigate through the choppy waters of uncertainty and global competition. History is also useful as a lesson. It illuminates the mistakes and follies of the past and guides the nation away from similar dangers. Here real and imagined deprivation, actual enemies and fantasized figures of evil run parallel with the reality of national achievement. One lends credence to the other. For the success of nation-building is never only about building so many public housing estates, or raising the people from the mire of poverty and unemployment; it is also evident in the triumphant great fight against enemies. For the Singapore State, this romantic narrative allows it to live parasitically on the national community. Peopled with heroic leaders and evil antagonists, the story of national struggle transforms the State into 'a being unto itself, animated with a will and a mind of its own'.[4]

Speaking generally then, national culture works by grafting the state to the national community. And it could not have done this without creating an imaginary place filled with wishes of how things should be, and figures of the evil Other conspiring to thwart them. The nation's Other is not all out to do us in, however; it is in truth remarkably double-faced. While it poses real or imaginary danger, the ghost of alterity also shows what the nation can achieve and, in so doing, presents itself as an object of emulation and cultural adoration. In Singapore, the PAP elite has come to identify the object in the dynamic world of the Chinese students and labour unions and, of course, in the West. Indeed the West is the perfect embodiment of things that the State warns people about, yet it also offers what Singapore most anxiously longs to acquire. If the West is the place of bodily indulgence and moral decadence that can infect our society and culture especially in these globalizing times, it is also the source of foreign capital, technology and, not least, designer goods of lifestyle consumption. For the Singapore State, the dilemma may well be phrased this way: what is to be done when the West matters so much to our prosperity and place in the modern world, and yet poses dangers of moral decay and cultural regression?

The trouble however is that it is sometimes hard to know exactly who the West is and what are its 'real intentions' towards us. Over the decades since Singapore's self-government in 1959, the West has taken on different meanings just as it has been garbed in garments of contrasting apparitions. These meanings and apparitions make the Western Other a figure of excessive imagination, something that comes alive in the busy traffic between fear and adoration, open repulsion and secret longing. To see the West as dream-like as I do is to be made aware of its rich significance and devious transformations as it operates in the 'dreaming subject'. And as in a dream, the imagining of the

West pits conscious thoughts against unconscious, forbidden wishes. As what is secretly wished for is expressed in clearly articulated forms, as the forbidden gives way to what is allowed, the West is fashioned in various shapes of displacement. Let us begin with the classic deployment of the West as 'dream-work' in the 1950s and 1960s: the official campaign against 'yellow culture'.

Yellow culture

The term 'yellow culture', a literal rendering of the Chinese phrase *huangse wenhua*, refers to cultural products like pornography, literature of love and romance, and pulp fiction of crime and fantasy, as well as the hedonistic and apathetic behaviour they inspire. The term also generally describes activities that sap energy and encourage masturbatory indulgence, making one oblivious to social commitment and perhaps revolutionary aims. Thus a person given over to the pleasure of pornography suffers corporeally and morally: the sallow, pallid face of a wasted body mirrors a selfish and an indolent spirit. Lee Kuan Yew has traced the idea to Mao's China:

> 'Yellow culture' was a literal translation of the Mandarin phrase for the decadent and degenerate behaviour that had brought China to its knees in the 19th century: gambling, opium-smoking, pornography, multiple wives and concubines, the selling of daughters into prostitution, corruption and nepotism. This aversion to 'yellow culture' has been imported by teachers from China, who infused into our students and their parents the spirit of national revival that was evident in every chapter of the textbooks they brought with them, whether on literature, history or geography. And it was reinforced by articles of left-wing Chinese newspaper journalists enthralled by the glowing reports of a clean, honest, dynamic, revolutionary China.[5]

In truth, the term 'yellow culture' was already in currency in Malaya and Singapore after the Second World War. During the twilight of British Rule, the colonial administration had been itself keen to promote a healthy, communal culture in order to 'fashion a "Malayan nation", infused with patriotic spirit'.[6] The form of culture and patriotic spirit was to fire up the country in the fight against Communist insurgents, and prepare for eventual independence. In any case, the search for a Malayan national culture would include censoring newspapers acting as mouthpieces of the Communists or the Kuomintang

(KMT) as well as Malay literature of excessive communal zeal. Hollywood cinema was also taken to task. The buxom sexuality of Jane Russell or Ava Gardner, and the glamour of crime depicted by Humphrey Bogart and Richard Widmark were thought to pose a threat to social order and the colonial authority. Hollywood was 'culturally questionable'.[7]

In a moment of political irony, 'the colonial policing of public taste' found a sympathetic ally in the 'anti-yellow culture' sentiment among Malayan radicals of various types.[8] The upshot was to broadly politicize the 'anti-yellow culture' campaign, turning it into a mixed bag of the progressive agendas. It strove for *merderka* or national independence, anti-colonialism and 'Asian socialism', while resisting individualism, materialism and the moral decadence of the culture of the colonial West. As the struggle for independence gathered pace, the West that so loudly featured in Hollywood films and other unhealthy imports became the common object of political and cultural struggle.

Anti-yellow culture in Nantah

It was in this heady ideological atmosphere that the newly elected PAP government found itself. Among its first initiatives, Lee recalls, were the street-cleaning drives and outlawing of 'secret societies, striptease shows, pin-table saloons and decadent songs'.[9] But there is a great deal of the 'copycat' – the word is Lee's – in the government actions. The most passionate 'anti-yellow culture' campaign took place elsewhere in the vibrant world of Chinese-educated youth, particularly on the campus of the newly formed Nanyang University (Nantah in the Chinese rendering).

Nanyang University was founded in 1953 to provide tertiary education for the thousands of graduates from the Chinese high schools in Malaya and Southeast Asia. For these students, finishing a six-year high school course had given them the cultural pedigree and modernist sensibility of a 'Chinese education', but they faced a bleak economic future. The English-based Malaya University, the public service and foreign firms all preferred graduates from the English schools. Nantah was built to redress the disadvantages facing the Chinese school students. In the Chinese tradition of communal effort, money was raised from Chinese millionaire philanthropists led by rubber tycoon Tan Lark Sye who himself donated $5 million, as well as from a massive funding drive by hawkers, rickshaw drivers, shopkeepers and dance-hall hostesses who contributed their daily earnings. Built on 500 acres of old rubber plantation, Nantah was to be the first Chinese university in the region and a triumphant symbol of Chinese education in Malaya and Southeast Asia.[10] Nantah came to shoulder considerable idealism and cultural ambitions of the Chinese community in Singapore. And as usual with such things in Malaya, it immediately attracted the labelling of 'Communist infiltration' by the government and conservative quarters. Thus one commentator has spoken of the students as being 'recruited into [Communist] cells and could agitate efficiently for political changes in the curriculum'; but even he admits that the 'students had genuine grievances' under colonialism and later independent Singapore.[11]

The comments came from the pen of Michael Thorpe, who went to Nanyang University in February 1962 as a lecturer in the department of modern languages and literature. His account, notwithstanding its imperturbable British views, gives a fascinating picture of student politics on the campus. Brought up on a diet of the modernist literature of the May Fourth, the young men and women who entered Nantah were no effete products of Confucian subservience and rote learning. 'From the onset,' Thorpe writes, 'Nanyang was doomed to become an instrument in the power-struggle between a west leaning meritocracy and those who, educated in the closed Chinese system, looked only to China as their legitimate parent.'[12] As Thorpe began his duties in Nantah, he was, quite in spite of himself, deeply moved by the students' discipline and the collective labour he witnessed.

They were helping to construct the students' union building – an answer to the Chinese traditional architecture of the library with 'columned portico and sweeping green-tiled roofs, dragons breathing at their corners'; Thorpe describes the hectic scene:

> [The student union building] was built on a defensible hill, itself largely an artificial product, its top flattened, its slopes graded and smoothed with baskets of soil and rubble borne by ant-like battalion of students. Slim girls toiled on the slopes under their own weight of earth. At weekends, while this 'beautification project' continued, revolution songs poured from the Tannoy loudspeakers erected on the slopes. Students were joined by the 'masses', with whom they believed themselves identified, in white shirts or tunics and loose black pyjamas:
>
>> These are girls like beautiful flowers,
>> Boys with strong bodies and open minds,
>> To build our new China.
>> We are happily sweating and working together . . .

So it went in the blaring song all would know, 'My Mother
Land'.[13]

'The organization, purpose and effort were, indeed, such as build
mountains', Thorpe surmises.[14]

To observers like Thorpe, the 'formidable Chinese application'
was impressive if a little alarming because it was accompanied by
much moral and ideological gravity, the stuff of revolution no less.
The qualities stood in sharp contrast to the largely Western-inspired
'yellow culture'. *The Sparks*, a magazine of the students' English
Society, chastised the newly formed Singapore government for failing
to stamp out 'bad films and immoral literature' and other evils:

> These [*sic*] yellow culture demoralises the youngsters who were
> at the impressionable age. Having been exposed to the influ-
> ence of such poisonous thought, they become gangsters, thugs
> or morally corrupted people. Students give up their studies and
> roam about the streets in parties of three or four. Gangsterism and
> juvenile delinquency are the direct results of such sexy pictures
> in which robbers, smugglers and rascals become heroes; beastly
> and barbaric acts are described as courageous, and perverted sex-
> ual behaviour is painted as the sole and normal life of men and
> women.

Young people turning to gangsterism and juvenile delinquency do not
make a stand for national struggle, and the 'colonialists' are to be
blamed for that:

> The morally corrupted people are trouble-makers and parasites
> of society. They cause a loss of man-assets to Singapore which is
> badly in need of them for nation building. We learn from history
> that it is the colonialists who were the patrons of yellow culture,
> who caused us to fall into the cultural traps set by them. They
> were self-interested; they tried to prolong their control over the
> people.[15]

A healthy body for nation-building, bravely warring against the
corruption of Western values: it could have easily come from the
new PAP government. For the government had itself made 'anti-
yellow culture' a part of the strategy for mass mobilization. As for
the students, their protest was more precisely over what they saw
as the half-hearted effort of the government in letting too many

'unhealthy' films, books and magazines pass through the censor's net. Nevertheless in targeting 'Western values' the students and the government did make a company of sorts. Both held a view of culture's real effects, and of the need to nurture healthy, rigorous Asian bodies for nation-building. In their uneasy alliance, they wanted to make the struggle against 'yellow culture' a central aspect of the wider, barely formulated project of a 'democratic-socialist' Malaya. At any rate, faced with the task of administration and in preparation for taking power from the British, the PAP government was to give the struggle an increasingly pragmatic tone.

Thus in 1962 Lee, then Prime Minister of the self-government, drew a lesson from what he saw during his six-week tour of Asia and Africa:

> I have enumerated in several of my talks of what I consider to be the three basic essentials for successful transformation of any society. First, a determined leadership, an effective determined leadership, two, an administration which is efficient, and three, social discipline. If you don't have those three, nothing will be achieved. . . .
>
> Where the social discipline is less, the progress is slow . . . If you don't get social discipline, everyone does what he likes to do, or will not bustle about what he is told to do. And that becomes the whole momentum [of decline].[16]

Here, as elsewhere, 'social discipline' is for Lee a dense phrase that brings to mind the ideas of bodily stamina, a 'rugged' national culture, and individual sacrifice for the good of the national community.

The body has always been Lee's chosen metaphor for depicting the rigour of the social and economic 'performance' of Singapore. As an idea of personal regime, 'social discipline' reins in human desires and 'excessive' pleasure-taking, fashioning as it does a prophylactic against the seduction of 'yellow culture'. It is an idea that would have nothing of the Western bourgeois view on the sanctity of the individual. In principle and often in policies, sexuality and marital choice lie within the locus of State intervention because they are plainly of 'national interest'. 'Yellow culture' as the cause of social mayhem thus proved to be a spellbinding idea. For that, the new PAP government was prepared to ignore the unsavoury association with Mao's China and the Communist-leaning left. Emboldened by spectacular electoral success, and with independence almost in its grasp, the PAP government refined the tenor and logistics of the 'anti-yellow culture' campaign. And it did this, predictably enough, by giving the idea a new twist while seeking out the enemy lurking threateningly at the gate.

Sarong culture, 'yellow culture'

On 17 November 1960, little more than a year after the PAP took power as a self-government, the newly appointed Johore Professor of English, D. J. Enright, delivered his inaugural lecture at the then University of Malaya.[17] The subject was 'Robert Graves and the Decline of Modernism', and in the lecture he made, he thought befittingly, 'a few topical remarks on culture, its equivocal nature and the acquisition or creation of it'.[18] But what he said evidently irked the government, as the local English daily the *Straits Times* reported the following day:

> A British poet, author and critic, Professor D. J. Enright, tonight fired another salvo at Government-sponsored attempts to create culture, and stressed that the most important thing for the two Malayan territories to do at present was to remain 'culturally open'.
>
> He said that authority must leave the people to fight their own battles, especially their personal battles. And he maintained that culture 'is something personal'.
>
> Mr. Enright also deplored the Singapore Government's ban on juke-boxes, spoke of the futility of instituting a 'sarong culture complete with pantun competitions', and suggested that every culture contained a trace of 'yellow'.
>
> He made these points while giving his inaugural lecture at the University.
>
> He first attacked the Singapore Government's attempts to create a Malayan culture and destroy 'yellow culture' in an article in the latest issue of the *Malayan Undergrad*, the mouthpiece of the University Students' Union.
>
> In his speech tonight, Professor Enright held that culture was built up by 'people listening to music and composing it, reading books and writing them, looking at pictures and painting them, and observing life and living it'.
>
> Using the word 'culture' in its widest sense, he said, the cultures of the Old World could be described as 'extremely cultural in the sense of being very distinctive, idiosyncratic, very different one from another'.

He added: 'Today the most distinctive national cultures are those which involve cannibalism, head-shrinking, or other forms of human sacrifice.

These days "national culture" is chiefly something for the tourists from abroad – the real life of the country goes on somewhere else.'

After this belittling of culture in the anthropological sense, according to the reporter, Professor Enright then proceeded to cast doubt on the whole exercise of making a 'national culture'. Culture should at best be left to itself and the people who live it, he suggested:

> Professor Enright said that to institute 'a sarong culture, complete with pantun competitions and so forth' at this point would be as ridiculous as 'to bring back the maypole and the morris (sic) dancers in England just because the present monarch happens to be called Elizabeth.'
>
> He added: 'The important thing for Singapore and Malaya is to remain culturally open.'
>
> 'Who can decide in advance which seeds will fall on barren ground and which will grow?'
>
> Claiming that all culture contained a trace of 'yellow', Mr. Enright said: 'Art does not begin in a test-tube, it does not take its origin in good sentiments and clean-shaven upstanding young thoughts.'
>
> It begins, he said, 'where all the ladders start, in the foul rag-and-bone shop of the heart.'
>
> To obtain art, and build culture, he suggested the following method: 'Leave the people free to make their own mistakes, to suffer and to discover.'
>
> 'Authority must leave us to fight even that deadly battle over whether or not to enter a place of entertainment wherein lurks a juke-box, and whether or not to slip a coin into the machine.'
>
> Professor Enright warned that 'a totalitarian state affords the most civilized society one can have,' because 'temporarily, at any rate, its citizens are essentially dead.'[19]

Later in the morning, Enright received a call from the Ministry of Labour and Law requesting his presence that afternoon on a matter about his passport. With a touch of the postcolonial settling of scores, his reception by the Ministry was, in his words, 'more on line of that traditionally meted out by the stern white master to the offending native'.[20] He was asked, on being shown a copy of the *Straits Times*, whether he had indeed said those things reported. Brushing aside his defence, the Acting Minister of Labour and Law thrust a letter at him. The gist of the Acting Minister's unhappiness was that Enright had overstepped his role as university professor, and meddled in Singapore's domestic affairs. As an 'alien' he had misused the hospitality by entering the political arena. The Acting Minister's letter insisted in no uncertain terms that:

> Whether the Government is right or wrong in banning jukeboxes or whether it should or should not try to foster a Malayan culture is a matter for the citizens of this country to decide. We have no time for sneers by passing aliens about the futility of 'sarong culture complete with pantun competitions' particularly when it comes from beatnik professors.
>
> This is to inform you that should you again wander from the bounds of your work for which you were granted entry into the country, then your professional visit pass will be cancelled as in all such cases. You are being paid handsomely to do the job which you are presumably qualified to do, and not to enter into the field of local politics which you are unqualified to participate in. You would do well to leave such matters to local citizens. It is their business to solve these problems as they think fit. They have to live and die in this country. You will be packing your bags and seeking green pastures elsewhere if your gratuitous advice on these matters should land us in a mess.

Lest Enright did not get the point, the letter ended with a reminder:

> The days are gone when birds of passage from Europe or elsewhere used to make it a habit of participating from their superman heights of European civilization.
>
> If you bear this in mind your stay in this country may be mutually profitable.

Of the whole affair, Enright might be said to have been a trifle naïve politically. Perhaps he did not think that a few words on national culture, casual preliminaries to the main text on Robert Graves and his poems, would have caused trouble with the authorities. Newly arrived in Singapore, he had perhaps yet to catch on to the local sensitivities, and the intricacies of national politics at the verge of independence from colonial rule. In any event, faced with the task of

building a national culture and challenges from the radicals within
the party, the PAP government could hardly let the incident go. The
'criticism', coming as it did from a white man speaking from the
prestigious height of the only English university in the colony, was too
significant to be ignored. The damning of the 'national culture' by a
liberal cosmopolitan, and the suggestion that juke-boxes, pornography
and other designs of 'yellow culture' should be left to individual choice
were provocative stuff indeed.

Enright's remissness, we might say, lay in underestimating the
seriousness of what 'national culture' meant to the new government.
He was a little myopic about what he himself symbolized in the heady
atmosphere of anti-colonialism and surging Asian nationalism. Having
taught in universities in Egypt, Japan, Thailand and other places, the
scholar-poet had probably found the making of 'sarong culture' too
parochial for his liking; but there is another thing.

Speaking of the twilight years of colonialism, it is hard not to make
much of the 'whiteness' of the Professor of English. For the Acting
Minister of Labour and Law, the Professor was wilfully ignorant
of the new world and still took for granted the privileges of his
'race'. 'Whiteness' fatefully aligned Enright to the colonial West. If
we can read the Acting Minister's mind, Enright and people like
him increasingly came to represent the old, tired European power now
facing insurgent wars and anti-colonial movements everywhere: he was
the feeble body that the 'anti-yellow culture' campaign had singled
out. Looking at it this way, the 'Enright Affair' looks increasingly like
an attempt by a postcolonial regime to redraw the contour of power
and, in the process, to fashion a new Asian identity at the dawn of a
new world.[21]

The West as dream-work

And it is a new world in which differences of 'us' and 'them', New
Asia and the waning West and, in the widest sense, of the young, virile
body and masturbatory waste were being marked. As these differences
rose from the euphoria of anti-colonial struggle, they were also allied
with urgently practical enterprises, as evident in the collective labour
of the students in Nantah under the scorching tropical sun. The West
is simply the silent, haughty Other that threatens to foil the Asian
ambition. With so much at stake, no wonder that the fight against
'Western values' has taken on an impetuous, excessive quality. The
West, feeble and corrupting one moment, imperiously domineering
another, is a potent symbol in the dream of a socialist Singapore.

There is, however, another side to the virulent accusation of the West. To go back to the beginning, we may ask: what did Lee Kuan Yew and his PAP colleagues want in these volatile years? From the Chinese-educated left, they wanted to harness their commitment and organizational skill, their working-class support, their hard work and incorruptibility. With these they would fashion the PAP as the only viable political force to which the British could safely transfer power. What they wished from the West was more ambiguous – even though they would at first appear to be after a total rejection of its cultural influences. Despite its socialist aspirations the PAP vision of development had included (as it still does today) reliance on Western capital. The tough policies – from disciplining the labour movement to creating social stability – were for creating a favourable investment climate and firming up the political and economic ties with Washington, London, Tokyo and other centres of the capitalist world. The State has never disguised the fact that it wanted a capitalist economy for Singapore, even as it attacks the West for its historical sins and moral weaknesses. All this makes the 'West' a complex and uncertain idea indeed, one that carries some interesting psychological subplots.

When we say that the West was the elegiac Other of Singapore, it is to suggest that the nation did not find its new identity only by obsessively thinking about itself, about its tragic history and plan for the future, even though it did a great deal of that. What the nation also needed was an idea which would show up its best qualities and achievements and, at the same time, light up the sad scenario should it give in to its enemies and to the dangers and corruptions they brought. It was an idea that exemplified the things it desired as well as those it disavowed. And there was no mystery what the embodiments of such an idea were.

The hardy Chinese bodies bravely facing the police in the strikes, and young students toiling under the sun in Nanyang University were for the PAP the classic object of desire and fear. Their strength and rigour had to be made use of. But these men and women were of a different ideology; they were a powerful political force that threatened to steal the political crown from the PAP. The West similarly figured in the calculation of needs. Just as the young, radical Chinese youth signalled political chaos, the West, personalized in the Johore Professor of English, pointed to moral and cultural dangers. The Western Other, of course, also promised delivery of desirable things, the importance of which the Asian seeker was hard put to admit. When constructed from the figure of the West, the Asian body is split into a set of twins of dual qualities: one dynamic, one feeble; one making for aggressive pointing of fingers ('You in the West are weak and decadent and out to corrupt us!'), one with a barely disguised need for the offerings of the West. The West is clearly more than those things that the 'anti-yellow culture' vanguards so passionately denounced. In the Asian imagining, the West comes to take on twisted, dreamy qualities.

For the master of dream-analysis the way we imagine all manner of subjects has to do with our past, and the conditions of our life. Freud is luminously clear that dreams are made in the busy traffic between unconscious wishes and socially grounded thoughts. Dreams, in his famous formulation, are like a border that joins, and separates, our inner and outer worlds.[22] We dream to act out wishes we are not socially allowed, like incestuous attractions or murderous rage against the father. Good dreams keep the dreamer peacefully asleep, and this is invariably about transcribing forbidden wishes into a safe and thus socially harmless form; Freud calls the process displacement.

What takes place in dreams is therefore a kind of negotiated peace between forbidden desires and the compulsive need to realize them. A convoluted affair indeed, dream-work energizes ideas and wishes, deploys images to *condense* and disguise them, and finally links these 'dream-contents' into a story line. Dreams are the condensation of many contrasting desires that appear in sleep in short and fragmentary forms, yet rich and multiple meanings are imbedded in them.

With displacement and condensation, Freud renders dreams into fluid and uncertain processes indeed. Both social in origin and lodged in the unconscious, dream-work cannot resort to crude equivalences of the 'lake or tunnel symbolizes the vagina' sort. The lake may stand for the vagina and is thus a disguised desire for sex, but something else is always 'condensed' in it so that the water may 'disguise' another longing. Freud would have us see meanings and their associated wishes as spreading out in 'associated paths', where one set of dream-thoughts is linked to another in an ever-expansive web of significations:

> Not only are the elements of the dream determined by the dream-thoughts many times over, but the individual dream-thoughts are represented in the dream by several elements. Associative paths lead from one element of the dream to several dream-thoughts, and from one dream-thought to several elements of the dream. Thus...a dream is constructed...by the whole mass of dream-thoughts being submitted to a sort of manipulative process in

which those elements which have the most numerous and strongest supports acquire the right of entry into the dream-content.[23]

Repulsion and needs

We need Freud's masterly dream-analysis to bring out the difficult ambitions of the 'anti-yellow culture' campaign. For one thing, it makes us see what happened with some coherence: the energetic Chinese body and the languidly feeble West are not so much opposite as lying coyly next to each other on the same network of dream-imagination. Each is the evocation of the other, just as each is a symbol that *condenses* different and contrasting passions. At the same time, it is clear that moral and physical dangers really do not explain the excessive virulence in the assault on the West. In the intersecting of dreams and repulsion, the passion of the PAP leaders spoke abundantly of the need for the rich offerings from the West that are always seemingly out of reach. In this sense, making it into a figure of cultural evil, an object of carnal weakness, looks increasingly like a *displacement* of longing to which the postcolonial Asian subject could not admit. Displacing a figure of desire with one of repulsion made particular sense in the heady days of decolonization. Only in the wakeful state of a fervent anti-colonial stance could the PAP view independence from the British with unqualified confidence, and Singapore's reliance on the West with sneering forgetfulness.

Indeed if decolonization was now inevitable, what happened in Algeria, Congo, Mozambique and, closer to home, Indochina would remind Asian nationalists everywhere that it was far from being a straightforward affair. The European powers' willingness to let go of their possessions could not be taken for granted. For Singapore, added to this is the fact of the crucial economic and military roles of the departing British – and of the United States.

At the time of Singapore's self-government in 1959, the British were the economic bloodline. By July 1967, two years after Singapore independence, British military spending amounted to 20 per cent of the GDP and employed about 10 per cent of the workforce.[24] Singapore sorely 'needed the British bases', Lee writes; 'if they were closed 44,000 workers would lose their jobs and the island would be defenceless'.[25] What Lee points to were two issues that continue to test Singapore's viability to this day: the economy and security. These are closely intertwined, as Singapore's location in Southeast Asia and its significant entrepôt trade puts a high premium on regional stability. In the volatile conditions of the 1960s and 1970s, such concern invariably translated into British, and after the withdrawal from east of Suez, American military presence in Asia. Asian leaders might wish to protest about continuing Western domination, Lee told his audience during a state visit to New Zealand, but the 'removal of military bases' put Asian 'national interest...in jeopardy'.[26] Indonesian opposition to the formation of Malaysia during *Konfrontasi* (1963–65) and the Vietnam War made urgent priorities of the defence arrangements with the United States and the Commonwealth (specifically Britain, Australia and New Zealand), and the building of a self-defence force. In regard to the latter, after independence in 1965 Singapore sought assistance from the non-aligned nations in line with its new position in decolonized Asia and Africa. It approached India and Egypt but the assistance was never forthcoming. Lee was especially disappointed when Egypt's strongman Nasser, whom he knew as a personal friend, 'opted out'.[27] In any event, Israel became the key architect of the Singapore Self-Defence Force, and Israeli advisers arrived 'disguised as Mexicans' so as not to upset the Muslim neighbours, especially Malaysia across the Causeway.

Western military presence was to have huge economic benefits as well; here is how Lee assesses America's venture in Vietnam:

> Although American intervention failed in Vietnam, it bought time for the rest of Southeast Asia. In 1965, when the US military moved massively into South Vietnam, Thailand, Malaysia and the Philippines faced internal threats from armed communist insurgencies and the communist underground was still active in Singapore. Indonesia, in the throes of a failed communist coup, was waging *konfrontsi*, an undeclared war against Malaysia and Singapore. The Philippines was claiming Sabah in East Malaysia. Standards of living were low and economic growth slow.

Quite simply, Singapore – together with most countries in Southeast Asia – benefited economically and militarily from American *imperium*.

> America's action enabled non-communist Southeast Asia to put their own houses in order. By 1975 they were in better shape to stand up to the communists. Had there been no US intervention, the will of these countries to resist them would have melted and Southeast Asia would most likely have gone communist. The prosperous emerging market economies of Asean were nurtured during the Vietnam War years.[28]

Occidentalism

When Edward Said describes the European imagining of the Orient as a 'collective daydream', he points to a terrain, an actual geography. The genius is that this geography does not have to refer to real places, but is filled with cultural myths, racial stereotypes and real institutions of power. The idea of the Orient – Orientalism – is at once a discourse and more:

> a distribution of geopolitical awareness into aesthetic, scholarly, economic, sociological, historical, and philological texts; it is an elaboration not only of a basic geographical distinction (the world made up of unequal halves, Orient and Occident) but also a whole series of 'interests' which, by such means as scholarly discovery, philological reconstruction, psychological analysis, landscape and sociological description, it not only creates but also maintains . . .[29]

Like dream, the imagining of the Orient has a history just as it has come out of the exchanges of powers and wishes. To speak of Orientalism as dream-work is to see the way it *condenses* all the desires in the European longing, and displaces forbidden wishes with others that can only show themselves in the light of day. The result is that, as the European eye turned to the East, it found native savages too prone to go on cannibalistic binges, but living in the natural state they could still be imagined as noble and spiritual. As for condensation, the Orient is that single, fantastical place 'over there' that promised to deliver all the things and ideas that imperialist ardour conjured up. A good deal of Orientalism may be the fiction of an over-excited imagination, but the effects and power were very real; it literally sent men off to the remotest places to pluck the fruits of imperialism – sexual pleasures, passive natives waiting to be saved and enslaved, El Dorado's gold, fame of exploration and more.

We might think of the 'anti-yellow culture' campaign with its stand against the West as a reversal of Orientalism: Occidentalism perhaps. The excessive passion traditionally used to inscribe the Orient is now reflected in the endeavours of the postcolonial subject. With Occidentalism the shoe is clearly on the other foot, so to speak. The West is now fashioned into the silent, passive figure of alterity. In this vein, the painful derogation of Professor Enright in the Acting Minister's office sounds increasingly like poetic justice at a time when world political fortunes were dramatically changing. And not only there, but in the Nantah campus too, the attack on 'yellow culture', the 'unreasonable' casting of blame, the shifting of moral culpability to the other side, and the perception of the Other as too feeble to answer back all remoulded the West into a potent symbol of dreams.

Yet all the while we suspect that the systematic divide between the West and an Asian state like Singapore simply cannot be. And we need Lee's pragmatic view of the Vietnam War to remind us of their intimate connections. For the 'anti-yellow culture' campaign, like Orientalism, is a sign of a subject's deep entanglement with the Other, an entanglement that invests in the Other such wondrous and yet abhorrent cultural and moral qualities. When the West is also what we welcome on our shores, it becomes an unmistakable symptom of displacement when it is 'fixed' as a singular sign of moral failure. The idea of the West as the evil Other hides its true usefulness, casting the unbearable longing for it to the dark of postcolonial ideological correctness. The struggle against 'yellow culture' is thus two stories in a single narrative of desire: one of rejection, one of desire; one of need and longing, one of fantasy of freedom and independence.

Asian Values

The time from 1965, the year of independence, to 2000 marked Singapore's progress 'from Third World to First', as Lee called the second volume of his autobiography. The economic success was largely driven by foreign investment, and Singapore contributed strike-free, disciplined labour, tax-holidays and other state incentives. The main sources of capital were, and still are, the United States, Europe and Japan. From the early labour-intensive industries to the current high-technology manufacturing and value-added services, the reality of foreign capital and technology in the Singapore economy, we are tempted to say, mocks the 'socialist fantasy' of heroic self-reliance and collective labour we have witnessed in the Nantah campus and the street-cleaning campaign led by PAP leaders.

That, however, was a time of different passions and practical concerns. Singapore as a First World nation signals something else: material prosperity, a culture of modernity and, of course, new relations with Washington, London, Geneva and Tokyo. Internationally, it also means fresh obligations (Singapore had a modest contribution to UN peace-keeping in Cambodia, and to the coalition forces in the 1991 Gulf War and the war in Iraq), and intimacies with nations and societies whose cultures and aspirations do not always resonate with one's own. Compared with the 1960s, the West is now an animal of a different shade, inciting dreams of a different sort. It says much of the globalizing world that the 'evil West' still presents the compelling figure of alterity, as it continues to mirror Singapore's own uncertainty just as it is eagerly seized upon to inscribe the country's new 'Asian identity'.

Nonetheless the 'anti-West' stance this time round is incited by moral panic of a different kind. It takes place in the context of close diplomatic dealings, at a time when top Singapore leaders enjoy a warm reception in the White House and Downing Street. Singapore's quarrel with the Western media, not to mention the hanging of Westerners found guilty of drug charges, has not impeded the flow of capital and smooth business dealings. In the post-Cold War geopolitics, the use of the West as a figure of alterity has to be a different undertaking from that of the past. If the assault on 'yellow culture' sounded like so much moral 'breast beating', boisterously disavowing the need for the West, Singapore's current international posture seems to dimly signal the opposite. The derision of the West this time is moved by the feverish need to affirm the island-state's evident achievements, and heralds its arrival in the First World of nations. In the State discourse on Asian Values, we are to hear the clamouring for a fresh assessment of the world and Singapore's new international position.

Given the vast and uneven literature on 'Asian Values', it helps to separate the sublime from the ridiculous. Michael Barr sensibly writes:

> The primary tactical premise of the 'Asian values' argument is one of cultural relativism: that many of the hegemonic political, social and cultural norms of the late twentieth century are western, rather than universal, norms and no more legitimate than alternative norms that could be considered 'Asian'.[30]

The proponents of Asian Values are ultra-sensitive to the criticism of Asian failings – on human rights, control of the media, lack of environmental standards and so on – and keen to label it Western cultural imperialism. They highlight national sovereignty, and also what they see as the realities of economic growth and cultural cohesion in Asian societies which have largely escaped the alienating effects of industrial capitalism. It is in short a near-hysterical voice of 'Asianism'. In Singapore there are many fastidious texts that loudly proclaim this; but, for brevity's sake, one will here do.

On 14 December 1993, the English daily *Straits Times* carried a piece called '10 Values that Help East Asia's Economic Progress, Prosperity' by Professor Tommy Koh.[31] The Professor was Singapore's former ambassador to the United States, and is currently chairperson of the Arts Council of Singapore and Director of the think-tank Centre for Policy Studies. Asian Values, to his mind, range from group orientation ('East Asians do not believe in the extreme form of individualism') to love of education ('Asian mothers would make any sacrifice to help their children excel in school') and the innate virtues of hard work and frugality ('East Asians believe [in living] within their means [in contrast] to the Western addiction to consumption'). Rhetorically the text dances with the intricate steps of the tango: moving to embrace the elegiac values of Asia one moment, and coyly repelling those of the West in another. Each affirmation of Asian virtues corroborates the lack of them in the West and thus the need to keep them alive for Singapore's cultural and material survival.

In these moves, it is clear that the 'Asian collectivism' is meant to endorse the central agendas of the state. For in '10 Values' the collectivism that comes so naturally to Asians consists of a hierarchy of increasing demands and responsibilities, with the family at the lowest rung and the state at the apex. An individual 'is not an isolated being', Koh writes, but finds his destiny as 'a member of a nuclear and extended family, clan, neighbourhood, community, nation and state'. But, as we read on, the family quietly slips away and becomes a metonym of the state. It is of course love and the primordial kinship sentiment that render parental authority and children's submission as morally fair and ethically inevitable. In Professor Koh's celebration of Asian Values the reciprocity between parents and children, organically constituted and morally natural, also becomes, on another plane, the basis for the 'Asian version of a social contract between the people and the state'. In return for the services and social peace provided by the state, 'citizens are expected to be law-abiding, respect those in authority, work hard, save and motivate their children to learn and be self-reliant'. This social contract has a major material outcome: it makes for frictionless working of the economy:

> East Asians practice national team work. Unions and employers view each other as partners, not class enemies. Together, government, business and employees work cooperatively, for the good of the nation.

   This philosophy, combined with the ability to forge national consensus, is one of the secrets of the so-called East Asian development miracle.

   Meanwhile the West features in all this as the opposite of Asian endowment; it is a sign of exactly those things that will undo the New Asia. For once those values that make Singapore a place of cultural harmony and industrial peace have been identified as 'Asian', they do not, as a matter of logic, exist in the West. Thus the charming polarities: 'Unlike the Western society, where an individual puts his interests above all others, in Asian society the individual tries to balance his interests with those of family and society'; 'Most Asian governments do not pay unemployment benefits,... partly to avoid the Western disease of welfarism'; Asian frugality 'is better than the Western addiction to consumption, paying "on time" and living under a mountain of debt'. But the West does not even have to appear in this pairing of merits and faults; so that a statement like 'Asian mothers make sacrifices to help their children to excel in school' would silently point an accusing finger at the callousness of Western women too prone to abandon children to their uncertain fate (in loveless marriages, and in the crime-ridden, poorly funded public schools, presumably).

   In the past, commentators and colonial officials tended to greet the attack on Western materialism and moral decadence with a mixture of glee and disdain. Michael Thorpe's reaction to Nantah students is perhaps typical. By comparison, the assault on Western values this time was met with interesting, often inflated responses from various quarters. For the progressive circles, 'Asian Values' articulate the angst of the *fin de siècle*, as Singapore tries to consolidate its power and international standing untrammelled by local and foreign criticism. But there are also enthusiastic appraisals, some from very distinguished people indeed. Former German Chancellor Helmut Kohl, for example, was certain that 'American and European leaders have profited from... [Lee Kuan Yew's] analysis and explanation of Asian values'. To Margaret Thatcher, Lee 'was never wrong' in clarifying 'the issues of our times and how to tackle them'.[32] On the writings of another promoter of Asian Values, Singapore Foreign Service official Kishore Mahbubani, Professor Samuel P. Huntington of Harvard University writes, '[He] has an instinct for the jugular when it comes to identifying a critical issue...; [his writings] will make Asians and everyone else think better than they have.'[33]

From 'rugged individualism' to the 'rugged Asian body'

These comments reveal much of the seductive aura of Asian Values – for locals and the international audience alike. Perhaps Asian Values simply make for an enunciation of the time; what gives them credibility is arguably the 'East Asian Economic Miracle' in the two decades before the 1997 crash. Singapore was the rising star in East and Southeast Asia: in the period it often achieved double-digit growth, enabling it to have the highest GNP per capita in Asia after Japan. The causes for the boom were a mixture of strategic, international and local factors. Taking a broad sweep across the coastal capitalist states from South Korea to the eastern side of the Indian Ocean, Benedict Anderson has argued, the economic growth in the region took seed in the massive economic – and military – aid from the United States during the Cold War. As he describes, '[In Southeast Asia] Washington made every effort to create loyal, capitalistically prosperous, authoritarian and anti-Communist regimes. . . . Each disaster only encouraged Washington to put more muscle and money behind its allies. No world region received more "aid".'[34] This, together with other foreign capital and the economic drive of the ethnic Chinese who make up the majority of Singapore's population, is a more realistic view of the country's success story.

In any case, for its admirers, Singapore's achievements are its final credential as a modern capitalist state. These achievements make it easy to look at its unsavoury practices with a tolerant eye. There may even be something in the 'Asian approach' to society and the economy that the West can emulate. Nevertheless Asian Values are also for 'domestic consumption'. For the Singapore State, we quickly note, the idea of Asian Values is remarkably thin on geopolitical and material realities; what is striking is the fixation with 'culture'. And culture here is simply the inner essence that shapes individual behaviour; good culture leads predictably to good behaviour (and to a productive economy). The trick is to make 'good (Asian) culture' the existential stuff of each person and, collectively, core elements of national identity. As in the days of 'anti-yellow culture', there is talk of discipline and hard work if only because these values make economic and productivist sense. However, the market economy also calls for healthy self-interest, in which case individualism may not be altogether a bad thing when it drives people to compete and excel. Hence in the 1970s Lee Kuan Yew could even laud

Singaporeans as essentially individualistic achievers ... because as immigrants, they might have developed a keen self-centredness which motivated them to work hard in their struggle to survive. The same spirit was central to the dynamism of the economy as a whole. [It was argued] that the extended family system could be an obstacle to economic growth because it discouraged one who has to share the fruit of his or her labour with others in the family to strive harder.[35]

There has been a remarkable cultural shift and change of heart since then. The stories of 'anti-yellow culture' and Asian Values have to be retold in this light. What I have described – from Nantah student radicalism and the Enright Affair to the discourse on Asian Values – is really about a mode of entanglement, a map of uncertain relationships between Self and Other. If the West can be seen as the figure of dream-work, it is to be in debt to Freud's brilliant but simple insight: to imagine the Other is to imagine oneself. In this sense, as economic prosperity and international standing compel the Singapore State to – aggressively and excessively – 'speak', they even more urgently call for creating immunity from those real and imaginary things that have defined oneself. That is why 'Asian Values' make so much of 'culture' and its effects. The West has to be fashioned as a sign that forewarns the Asian State of the intolerable scenario: the breakdown of the 'community' and 'Asian identity', and with this the rupture of State power and authorities. Asian Values, quite simply, are a panacea, a wall of immunity against a world increasingly coming together. Western modernity and the goods of urban consumption are to be enjoyed, but the make-up of the Asian subject – respect for authority, regard for the community, love of the family and so on – must not be left by the wayside.

If Asian rigour has to be protected, it is not carnal strength – which in any case can be imported in the form of foreign labour from poorer Asian neighbours – but culture that is the concern. The ambivalence towards the West is primarily concerned with how to build a cap-italism that gels with Asian Values. All the exuberant, passionate announcements of East–West 'differences' lead to this point. When Asian Values are taken to equate with Confucianism, it creates the so-called 'Confucian capitalism', one void of competitive violence, industrial conflict and alienation of the sort that supposedly plagues the West.[36] With Confucian capitalism's culture, as with Professor Koh's '10 Asian Values', economic calculations and pursuit of profit do not go together with heartless market forces, but sit cheek by jowl with social harmony and moral considerations. The West is simply a warning of how vulnerable this project can be.

If economic arrival has reconfigured Singapore's relations with the international world, it also gives it a new condition from which to speak. This, of course, is the new postcolonial posture. Among other things, the posture gives way to a different imagining of the Other at a time when 'we' are increasingly tied to the rest of the world, a time when the Western Other is already residing in our home ground. The dream of the West, like the dream of incestuous lust, is always to be reminded of the irresistible but forbidden longing. In the past the West could be moulded as the opposite to the new, virile Asia; it is now channelled into a new trajectory of meanings in daydream's expansive network of significance. In daydream's river of connections, the tributaries of us and them, fear and longing, innocence and corruption threaten to break their banks. That is why the idea of Asian Values has a distinct sense of moral panic about it, one that invites responses that are, by their very nature, recurrent and excessive.

4   Pain, words, violence
The caning of Michael Fay

[F]or the person in pain, so incontestably and unnegotiably present is it that 'having pain' may come to be thought of as the most vibrant example of what it is to 'have certainty'... Thus pain comes unsharably into our midst as at once that which cannot be denied and that which cannot be confirmed.

Elaine Scarry, The Body in Pain

Judicial caning in Singapore

Like much of the penal legislation in Singapore, judicial caning has its historical roots in the criminal justice of England and colonial India. Stamford Raffles, who landed on Singapore in January 1819, had acquired the island on behalf of the East India Company with its headquarters in Calcutta. From 1819 to 1858 Singapore was administered more or less as an appendage of British India. The Indian Penal Code together with British common law was extended to cover Singapore. Throughout the period, corporal punishment was imposed on the offences of 'begging, pornography, garrotting, treason, and robbery with violence'.[1] In 1867 Singapore – together with Penang and Malacca – became Crown Colony of the Straits Settlement, and the new Straits Settlement Penal Code replaced the existing laws. The Straits Settlement Penal Code prescribed whipping for a number of crimes including aggravated forms of theft, house breaking, assault with intention to outrage modesty, second or subsequent offences relating to rape and prostitution. Singapore became a separate crown colony in 1946, and the penal code retained most of the previous provisions for whipping, but now pain was administered by caning. These provisions were inherited by the independent state of Singapore, which has since then expanded the use of judicial caning to a list of other crimes.

Currently sentence of caning is applied to three categories of offences. The first is violent crimes or crimes with serious risk to the public; these range from kidnapping and robbery to unlawful possession of firearms. The second category consists of disciplinary offences in prisons, detention centres and the armed forces as 'deterrence' against refusal to comply with orders. Finally under the category of miscellaneous offences are vandalism (to a maximum of 6 strokes of caning), drug trafficking (from 5 to 15 strokes of caning), import and unlawful discharge of fireworks, and immigration offences including entry without a visa, overstaying, and conveying and employing illegal immigrants.[2] In many of these cases, caning is mandatory in addition to prison terms.

For all such offences, caning is carried out with the kind of efficiency and administrative precision for which the Singapore State is renowned. The court cannot order caning for women and for men over 50; persons already sentenced to death cannot be caned. The maximum number of strokes that can be awarded in any one trial is 24 in the case of adults and 10 in the case of young offenders (legally defined as those between 7 and 16 years of age). For caning of young offenders a lighter cane is used. In response to foreign and local critics, the Prisons Department has attempted to 'set the facts straight', and provided details of how caning is carried out.[3]

The prisoner is taken to the caning room. There he is examined by a medical doctor to ensure that he is physically fit to receive punishment. He is stripped naked and made to bend over and then strapped at his ankles and wrists with leather cuffs to an H-shaped trestle. A protective pillow pad is placed above the buttocks over his kidneys. Under the watchful eyes of the superintendent of the prison, a medical doctor, and a prison officer who will keep count of the strokes, the caner, standing about 1.5 metres away and with the full swing of his body, strikes on the buttocks. The cane he uses is 1.2 metres long and 13 millimetres thick as prescribed by law. It has been soaked in water and treated with antiseptic so that it is supple enough to produce the effect of a lash when struck against the body. The point of contact on the body produces splitting of the skin and after the third stroke the buttocks will be covered in blood.

At the first landing of the cane, prisoners usually put up a struggle. After that, according to the prison official:

> their struggle lessens as they become weaker. At the end of the caning, those who receive more than three strokes will be in a

state of shock. . . . Many will pretend to faint but they cannot fool the prison medical officer [who is on hand to revive them].

After caning, the prisoner is examined by the doctor and has his wounds treated. The Prisons Department official explains:

> Our law only prescribes caning with a rattan cane, and not flogging with a whip or cat-o'-nine-tails. The punishment is also administered by a prison officer, and not necessarily by a martial arts expert. The officer meting out the punishment is, however, required to be fit physically. . . .
> Any prisoner found to be unfit will not be caned. Similarly, if the doctor is of the opinion that the prisoner is unable to continue to be caned, he will order it to be stopped immediately.[4]

And he assures the public, 'Caning does not cause "skin and flesh to fly" as alleged by critics'; but the wound will leave indelible marks on the body after healing, a source of shame throughout the rest of his life.[5]

Pain, text, meaning

Talking about the meaning of pain is like talking about starvation in Somalia as a discourse. Both risk committing an ethical sin. For the sufferers, like the recipient of Singapore's judicial caning, can pain be anything else but . . . pain? Considering the social and cultural meaning of pain or any instance of human suffering takes us away from the materiality of bodily experience. The significance of pain, one might say, is always encased by its excruciating certainty. Pain's corporal sensuality thus creates a closed, secret world of the sufferer who feels it self-possessively and alone, in his body. Pain, in the words of Elaine Scarry whose magisterial *The Body in Pain* we shall later turn to, is characterized by its 'unsharability', its 'resistance to language'.[6]

There is a profound irony in this, however. If pain is nothing but pain, and if my pain is undeniably and incommunicably mine, then this already makes for meaning of a kind. The sureness of carnal experience, the certainty of knowing through the body, fashions the view of pain as beyond social sharing. 'Pain as mine alone' also insidiously belittles the moral sympathy that, however difficult, momentarily joins the sufferer and non-sufferer, and their separate worlds.[7] Against this tenet, the punishing power must impose its will not only on the body, but also on what pain means. Power and the meaning of pain are intricately tied.

Judicial punishment adds further complications; for here delivery of violence – and incarceration and death – takes place within the interpretation of law. And interpretation of law is about the assiduous reading of legal texts: in this sense, text, meaning, violence form the crux of the working of law. This is the contention of Yale University Professor of Law and Legal History Robert M. Cover:

> Legal interpretive acts signal and occasion the imposition of violence upon others: A judge articulates her understanding of a text, and as result, somebody loses his freedom, his property, his children, even his life. Interpretations in law also constitute justifications for violence which has already occurred or about to occur. When interpreters have finished their work, they frequently leave behind victims whose lives have been torn apart by these organized, social practices of violence. *Neither legal interpretation nor the violence it occasions may be properly understood apart from one another.*[8]

Judicial punishment may be about the execution of pain or death, or the deprivation of freedom; it is also about the struggle over the final meanings of these punishments, so Professor Cover reminds us.

All these are at stake for judicial caning in Singapore: thus the ritual precision and formal procedures of caning, which make up something like a 'manual' of a technology for extracting pain with efficiency. They make sure that the painful punishment is carried out fairly according to law. And they impart the meaning that caning punishes by inflicting excruciating pain on those out to damage social peace. As the Singapore State is never tired of insisting, there can be no other meanings than this. However, as we shall see, when judicial caning was being applied to a foreigner, a citizen of the most powerful nation in the world, the significance could hardly be only what legal formalism would have it. Powerful international criticism was to raise important questions about human rights and Singapore's approach to law and order, and these questions invariably put caning beyond the framing as a means of delivering just and necessary punishment. In its heated responses to international critics, the State would emphasize that the use of judicial caning has been responsible for the country's low level of crime. Besides it is also a question of national sovereignty and the right to implement its own 'Asian approach' to crime and punishment. In this way, to use Professor Cover's phrasing, judicial caning must transform pain into sign, the punishing experience into text. In Singapore, what this sign and this text call up is the unwavering certitude of its actions as legitimate and for the good of society precisely at a time when such certitude is being brought into question.

All this came to the fore in 1994 when an American teenager was found guilty of vandalism charges and sentenced to be caned.

The caning of Michael Fay

On 18 September 1993, cars belonging to the wife of a high court judge and her neighbour were found sprayed with red paint while parked in Chatsworth Road, an exclusive residential area. After they made the report, the police began to link the damage to the recent spate of vandalism in the area. A day before, similar paint spraying was reported in a multi-storey car park along Cairnhill Road where six cars were damaged. About a week later, on 26 September, two cars parked in the neighbourhood were splattered with eggs and apparently smashed by kicking or hitting with a brick. In the following month, on 4 October, vandals pelted two cars in the same area with eggs and took away the rear licence plates.

Police detectives set out to investigate and, in the early hours of 6 October, put up an ambush near Chancery Court. After giving chase at 3 in the morning, the police arrested two young men: the son of a Thai diplomat, and a 16-year-old boy from Hong Kong, Shiu Chi Ho, a student at the Singapore American School. The Thai youth was released after his diplomatic status was verified. During a seven-hour interrogation, Shiu gave the names of those he claimed were responsible for the acts of vandalism. Later that morning, about two dozen police officers arrived at the Singapore American School and took five students away in police vans. The five arrested were two 15-year-old Malaysians (whose names could not be reported as they were under age), an Australian, Damien Kirchhoff, aged 16, and two Americans: Stephen Freehill, 16, and Michael Fay, 18. A Belgian boy and two Americans were taken in for questioning in a separate arrest.

Fay took the police to the twenty-first-floor apartment where he lived with his mother and stepfather in exclusive Regency Park. In his room police found Singapore flags, two 'Not for Hire' taxi signs, a 'No Exit' sign, a 'Smoking Strictly Prohibited' signboard and other similar items. Fay had apparently stolen these, except the flags and signboard which were given to him by a Swedish school mate who had since left the country. Fay being the oldest and the most serious offender became the prime target of police investigation. He eventually signed a full confession of his crimes after nine days in custody, while the others were released on bail. Among those charged were Fay, Freehill, Shiu, the two Malaysian boys, and the Australian boy Kirchhoff who later escaped bail by leaving the country with his family.

While Shiu decided to plead not guilty, Fay and one of the Malaysian boys admitted guilt in order to negotiate a plea bargain with the prosecutor. The Malaysians were eventually sentenced to two months in a boys' home by a juvenile court. The government now turned its attention on Fay, who faced 53 charges, mainly related to vandalism. He and his friends were held responsible for damaging 18 cars over a ten-day period. Fay finally pleaded guilty to two charges of vandalism, one of possessing stolen property and two counts of mischief. On 3 March 1994, Fay was sentenced to four months in jail, a $2,200 fine and six strokes of caning.

The sentencing made headline news worldwide and started a heated media debate especially in the United States. Calling the court ruling 'extreme', President Clinton asked the Singapore government to consider waiving the caning. Following the failure of his appeal, Fay through his lawyer asked Singapore's president for clemency. On 4 May, the government announced that it would recommend to the President that Fay's caning be reduced from six to four strokes, as a sign of respect for the American President. The news was conveyed to Fay by his lawyer at noon the following day. Soon after the lawyer left, Fay was informed that his sentence would be carried out that afternoon. He was flogged together with ten other prisoners. The sentence was swift and over in minutes. 'After the fourth and final stroke', a prison spokesperson told the press, 'Fay shook hands with his caner and insisted on walking back to his cell unaided. He wanted to act like a man.'[9]

International furore and local responses

The sparse description culled from media reports[10] barely conveys the heart-wrenching details and the virulent responses from various quarters. Fay's mother, Randy Chan, had desperately tried to seek reprieve for her son. She broke down after hearing that caning had been carried out, calling it 'torture'. Fay's father, George Fay, had advised his son to plead guilty in order to strike a bargain with the prosecution. Racked by his own culpability, he had wanted to trade places with Michael. 'If you want to penalise somebody, then penalise me, and I will take his place', he said.[11] Then there was Fay's own fate at the end of the rattan, so to speak, as he was strapped on to the trestle to receive his due.

After the Fay caning, readers of the English daily *Straits Times* were given news of support from the United States and England. The sentiment was very much of the 'when in Rome...' sort. '[He] is over there [in Singapore], he has to abide by the rules', ABC Television's Ted Koppel said.[12] Implicit approval then quickly gave way to the fantasy of 'let us have caning too'. Anne Diamond of the *Sun* injected solemnity to the British daily more renowned for its 'page-three girls' by suggesting: 'Singapore has an astonishingly low crime rate. I believe [caning] would work here if we allowed it. Yes, it's answering violence with violence – but what is working, for goodness' sake?' Michael Fay's punishment by a tough, crime-free Singapore was made to reflect on situations in New York and London. For those clamouring to bring in the cane, Singapore summons up the ills of the West all too evident in the inner city streets, the state schools and the 'public housing hell'. For a moment, its approach to law and order seems to be a viable one for the West as well. A reader wrote from Huntington Beach, California, 'America should be taking lessons from Singapore on how to prevent crime – hold the line – don't give in.' To columnist Russell Baker of the *New York Times*, the support of caning is a case of 'the collapse of American ego':

> Half the country is cheering for a foreign government to go ahead and beat an American, age 18, to bloody pulp because he vandalized cars and signs in Singapore. This was rotten of him, no doubt, and he is also sentenced to do jail time for it, but it says something about the collapse of American ego when half the country starts cheering for foreign powers to beat up our fellow citizens.

The protest over the caning tended similarly to overstate the case. For William Safire of the *New York Times*, 'This issue is not about degrees of harshness. It is a case of a state asserting an intolerable "right to torture."' Richard Cohen of the *Washington Post* reminded his readers of the 'terrific punishment' and asked rhetorically, 'Would truly harsh penalties turn New York into a Disney World with skyscrapers? Indeed, would these sorts of punishments reduce crime in the United States?' The United States government made plain its position in similar terms, 'We continue to believe that caning is an excessive penalty for a youthful non-violent offender who pleaded guilty to reparable crimes against private property', and President Clinton had condemned the punishment as 'disproportionate'.

As for local responses, the *Straits Times* received some 300 letters up to April 1994, a month before the caning was carried out. Practically all favoured the sentencing. In these letters, Fay was seen as an American teenager spoiled by the good life of a wealthy expatriate family and the lack of firm parental guidance. After her divorce, Mrs Randy Chan had remarried and come to Singapore with her husband, Marco Chan, who was the managing director of the US courier services giant Federal Express. At the American School, he was known as an 'under-achiever and trouble maker', 'consistently indolent and defiled [*sic*] class discipline'.[13] So his broken home and separation from his father could have offered the 'mitigating factors' in his behaviour. But in the reader's mind, Fay's misconduct should not be explained away. Instead the American teenager was turned into a figure that expressed the national feelings about law, punishment and the nature of government – in contrast to practices in the West.

Among the letters, there were firstly those that mimic ad nauseam the official line about the deterrence effect of harsh punishment. Typical perhaps is reader Mark Wang Yew Kong who wrote: 'The US liberal judicial system has led to an ill-disciplined, crime-ridden society. In contrast, Singapore's tough laws have helped to keep our country relatively crime-free.' To the grieving mother, Winston Chin Fook Min offered this counsel:

> I read with disbelief in the *Straits Times* quoting Mrs Randy Chan as having alleged that her son, Michael Fay, her husband and herself 'have been followed by as many as 100 intelligence people in Singapore since sentencing'....
>
> I think she is flattering herself and her family. She has, wittingly or unwittingly, blown up the case of a common cold into a flu epidemic.

For my university colleagues, Fay is not exactly Nelson Mandela: he advances no 'progressive causes' so his punishment is not a subject of moral curiosity. In the context, the sentiment of surgeon-turned-novelist Gopal Bartham is a stark rarity:

> Singaporeans should... reflect on whether caning for non-violent crimes is compatible with our aspirations to build a humane, caring society.

Are things more precious than human flesh? If they are, the man who claims 'I broke his head because he dented my Mercedes' will have justice on his side.

Other readers so closely identified themselves with the State as to suggest their complicity. Tin Keng Seng thought that the President must turn down Fay's clemency plea, because

> to revoke the sentence would smack of double standards and Singaporeans would soon question the rule of law here. The principle that the same law should apply to everyone will be doubted by Singaporeans. There must be uniformity in sentencing for offences of the same nature.

Another reader also thought that President Clinton, in seeking clemency for Fay, was 'interfering in our judicial system', and 'the psychological damage' suffered by a criminal was of no concern to law-abiding Singaporeans. He then turned the comment to himself and his fellow citizens: 'Everyone in Singapore is subject to the same laws. To expect us to make an exception would be tantamount to asking us to disregard our laws. How then are we supposed to punish similar offenders in future?' The sentiment seems to be that, if no one should escape from Singapore's draconian law, then we – both locals and foreigners – are in it together. When a moment of sympathy shows itself, it is not to commiserate with the plight of Fay and his mother, but to offer the only suitable advice – Fay should take it like a man:

> Michael Peter Fay had his fun. He had his cheap thrills. He must now take his punishment like a man – and not let his poor mother (she will feel that pain more than he does, I am sure) and his stepfather involve their President in the hope that he can save himself from receiving 'six of the best' on his butt.

There is in these letters something of the reprimand from 'fellow travellers'. Fay's possible reprieve seemed to arouse near envy ('Why should he be allowed to get away with it when we locals can't?'). Perhaps it is a reaction from people long living under harsh laws, who must cultivate heartlessness as a psychological adjustment. So it is with the readers' slavish aping of the State in the attack on foreign critics. Nationalist passions are by nature parochial; here the perceived deterrent effects of caning, the retribution Fay rightly received and, above all, the aggressive settling of a score with the West all seemingly galvanized the nation, uniting the society and the State against those who questioned Singapore's right to punish.

State violence and 'order and law'

In the death penalty, Walter Benjamin writes in 'Critique of Violence', the law is shown at its most problematic.[14] The legal system, of course, uses violence to turn its rules and regulations into law. Without this 'law-making violence' there is no law enforcement, and rules and regulations are merely words and texts. But the death penalty also constitutes what he calls 'law-preserving violence'. While 'law-making violence' gives legal texts their practical effects, 'law-preserving violence' eagerly patrols the borders dividing the lawful and the unlawful:

> For the function of violence in law-making is twofold, in the sense that law-making pursues as its end, with violence as the means, what is to be established as law, but at the moment of instatement does not dismiss violence; rather, at this very moment of law-making, it specifically establishes as law not an end unalloyed by violence, but one necessarily and intimately bound to it, under the title of power.[15]

Violence is a necessary means for law-making. Once established, Benjamin suggests, law does not 'dismiss violence', but becomes dependent on it. Indeed there is a general tendency in law to regard violence not only as a means, but an end in itself. 'Law-making is power-making, and, to that extent, an immediate manifestation of violence', Benjamin writes.[16] Professor Cover agrees. Violence enforces the legislated texts as rules; it is also a part of the 'normative universe' that gives the law, and thus the state it upholds, dignity and authority.[17] Violence gives meaning to the legal institution and then 'restructure[s] it in the light of that meaning'.[18]

These notions of state violence, we should perhaps note, are not about soliciting moral sympathy for its victims. More importantly, they make the point that the law cannot function without violence, and the state's right to take away a criminal's life or freedom is an intrinsic part of the processes of law. Without such violence there can be no law and less still the enforcement of it. Interpretations of legal texts and the formal rituals in the courtroom, Professor Cover contends, are indeed 'themselves implements of violence'.[19] And this is his powerful argument: judicial violence does not take place only in the prison; it begins in the beginning, in the law's 'interpretive character' and the 'meaning of the event in the community of shared values'. Legal interpretation is thus profoundly practical:

> The judicial word is a mandate for the deeds of others. Were that not the case, the practical objectives of the deliberative process could be achieved, if at all, only through more indirect and risky means. The context of a judicial utterance is institutional behavior in which others, occupying preexisting roles, can be expected to act, to implement, or otherwise to respond in a specified way to the judge's interpretation. Thus, the institutional context ties the language act of practical understanding to the physical acts of others in a predictable, though not logically necessary, way. *These interpretations, then, are not only 'practical,' they are, themselves, practices.*[20]

The words of law are, literally, words of warning as they are intrinsic to judicial practices and to the power of punishment. The other way is equally true. Legal punishment confirms the judgment of blame, a validation of the so many words spoken and texts referred to in the court. When the day of reckoning comes, violence applied on the socially recalcitrant comes across as nothing more than just desert.

Yet, as judicial violence is finally about meaning too, what happens in the courtroom ever so subtly brings in the values and discernment of the social world. This is not to suggest that legal interpretation is in any sense an arbitrary process, where meanings can be plucked out of a hat to suit the state's purposes. What is true is that the law and law-making have a history, just as the application of law has a social and political context. For Professor Sally Falk Moore of Harvard University, the law is as much about 'the distribution of specific procedures or concepts or rules' as about 'the kind of society in which [it] operates'.[21]

On the issue of the law and its social context, Professor Sally Falk Moore finds an unlikely ally in Lee Kuan Yew. In 1962 the then Prime Minister spoke to members of the Singapore Law Society where he outlined what he saw as the right spirit of law:

> In a settled and established society, law appears to be the precursor of order. Good laws lead to good order.... But the hard realities of keeping the peace between man and man and between authority and individual can be more accurately described if the phrase ['law

and order'] were inverted to 'order and law', for without order
the operation of law is impossible.

What the Singapore State needed, Lee explained, was powers of
detention that could put away political offenders and secret society
gangsters. Law would have substance and real effects only if state
order, and the power to protect it, were in place.

> [When] a state of increasing disorder and defiance of authority
> cannot be checked by the rules then existing, new and sometimes
> drastic rules have to be forged to maintain order: so that the
> law can continue to govern human relations. The alternative is to
> surrender order to chaos and anarchy.[22]

For Lee then, it is decidedly the case of good (state) order leads to good
law. The state through its right to punish creates the proper conditions
in which law can effectively operate; without these conditions the law
is but a collection of empty words and texts. In the land of lawlessness,
the impotence of state order is the impotence of the law to punish. In
this land too, the only effective meaning of law is that it should uphold
the power of the state and thus the peace and order of society.

   A British-trained lawyer, Lee would realize that putting 'order'
before 'law' could hardly measure up to the principle of habeas corpus
and the idea of individual liberty enshrined in two centuries of the
development of English law. But it was a volatile time in Singapore.
Conventional strict application of the laws of evidence for criminal
trials would let too many criminals go free. The power of the state has
to be enhanced to make the law work: this was Lee's argument. The
reversal of 'law and order' into 'order and law' is also to make the state
master of law, rather than its servant as in the liberal tradition. And
when judicial violence is used to uphold state order and its mechanism
of punishment, the tendency is to turn state power into an end in itself.
For Lee the defence is clear enough: without state order not only the
law but society too is impossible.

Law and meaning-making

When Lee spoke at the Law Society in 1962, Singapore had just
three years earlier won self-government. The ruling PAP faced serious
challenges from the left in the party. They opposed Singapore merging
with Malaya to form the Federation of Malaysia in a plan announced
in 1961. Already disfranchised after their release from detention

when PAP assumed power, the 'pro-Communists' believed – quite rightly – they would be vulnerable to suppression and 'security operations' by the Federal Government in Kuala Lumpur. With the violent labour strikes still fresh in the memory, Lee called for tough measures and effective means of controlling these and other 'subversives'. Drastic measures were needed to protect social peace as much as PAP's tenuous hold on power.

Lee's conception of law is thus practical, setting out how it should operate in the conditions of the time. As with most of his speeches, this is 'thinking aloud' to work out an important idea for himself and the audience. The 'order and law' argument gives the suppression of gangsterism and leftist subversives a philosophic gloss and social purpose. Rather than for upholding State power as Benjamin would see it, the idea is but a demonstration of the State's quickness to incarcerate and administer pain to criminals. Echoing Professor Cover, what Lee had attempted was precisely to show law as a meaning-making enterprise too. Thirty years on, in dramatically different circumstances, the constructive, meaning-giving process of law has not radically changed but is taken to a new height when the State finds itself having to cane a young citizen of the United States of America.

On the Fay caning, Lee would typically cut deep into the United States' approach to law and order. The United States 'dares not restrain or punish the individuals, forgiving them for whatever they had done', resulting in 'drugs, violence, unemployment and homelessness, all sorts of problems in society', he said.[23] Singapore does not fall into such weakness; '[it] metes out harsh punishments to deter criminals ... [the] alternative is to surrender order for chaos and anarchy'.[24] Others voiced similar sentiments. If they found the legal language too coldly formal, they would spice up their comments with a bit of the modish 'East–West cultural differences'. Here is the view of businessman Ho Kwon Ping:

> The Western cliché that it would be better for a guilty person to go free than to convict an innocent person is testimony to the importance of the individual. But an Asian perspective may well be that it is better that an innocent person be convicted if the common welfare is protected than for a guilty person to be free to inflict further harm on the community ...[25]

For Walter Woon, law lecturer at the National University of Singapore and a nominated Member of Parliament, caning gels with 'Asian culture'. 'No matter how harsh your punishment you are not going to get an orderly society unless the culture is in favour of order', he said.[26] When Fay's caning was reduced from six to four strokes after President Clinton's plea for clemency, the then Acting Prime Minister Lee Hsien Loong was moved to say, in a rare display of public humour for the man, that there was no reason why 'foreigners should be more thin skinned [compared with Asians]'.[27]

The comments relentlessly gave the same message about Singapore's approach to crime: that the harsh and efficient application of pain to the recalcitrant is an effective deterrent. Caning is nothing but pain-giving: nothing should obscure this purity of purpose and meaning. When the person to be caned is a foreigner, a citizen of the world's most powerful nation, the fact complicates but does not shift the ground of the argument.

For us, there are some crucial questions about state power and, particularly, about how the responses of the state – and its support-ers – help to give the law substance and potency. If meaning-making is also law-making, then the answers surely lie in the words and their usage. The aggressive talking back to the West in this and other in-stances is not simply for explaining the issues at hand; it is also for reconfiguring the significance of the law in newer circumstances. The 'order and law' this time no longer concerns itself with gangsters and radicals, but with social recalcitrance of a more transnational kind. Here the law and Singapore's position in the world became sorely tested.

As with the Bible or a pronouncement from the Vatican, we cannot think of the words of the Singapore State without thinking of the political force or institution that put them in its mouth, as it were. What is at stake with the Fay caning, quite apart from the punishment itself, is the way it puts Singapore up to international judgement about the country's standards of human rights and principles of law. In the cacophony of sentiments and responses, Fay's punishment is slyly turned around to say something about Singapore and the West. In remarkable disparity, Singapore's tough, more 'realistic' treatment of criminals is pitted against the Western emphasis on rehabilitation and repentance; Singapore's safe streets and communal cohesion are pitted against the urban crime and cultural malaise of the West. The brilliant accomplishment of one cannot but show up the pathetic failures of the other.

It becomes classic meaning-making when the argument, so trans-parently true for Singapore, hides some important facts. 'Painful punishment as deterrence' is put forward as the sole reason for judicial caning, so unimpeachable that there is no need to pay heed to the fact that, in Hong Kong, another Asian country based on a so-called Confucian heritage, crime rates have declined following the abolishment of the death sentence and caning, and caning for non-violent offences like vandalism ceased in Britain in 1948. In a similar vein, local lawyers I talked to would tell me that Singapore has acted within the law as the Constitution does not include a guarantee of freedom from torture or degrading punishment, and that it is not a party to the international convention on human rights. And is not the United States being hypocritical, they said, when it allows capital punishment by various means? They are, however, notably silent on another foundation of judicial punishment, the principle of proportionality – of 'punishment to fit the crime'. In the opinion of Firouzeh Bahrampour of Washington College of Law, 'while American capital punishment and Singaporean caning are both painful, they are not both disproportionate. Capital punishment is proportionate for such crimes as first degree murder and treason; however, caning is disproportionate punishment for property crimes.'[28] Caning may be effective and necessary punishment, but it fails the test of proportionality, Firouzeh Bahrampour suggests.

Perils of globalization

Not surprisingly, the belligerent anti-West, anti-American positions are not quite what they seem. Singapore's apparent tiff with the international media must not upset the economic and diplomatic relations with the most powerful nation and an important trading partner. In the end, the fervent 'cross-cultural misunderstanding' and wounded national pride led to nothing – except for Fay the prisoner. When the *New York Times* called for American corporations with investment in Singapore to pressure the Singapore government, Citicorp, IBM and DuPont refused; Exxon and Coca-Cola 'had no comment'. Business was too important to have it interfered with by a squabble with Singapore over an unruly teenager. Neither did the US State Department do more than ask for leniency. For Singapore, the 'quarrel with the West' was managed, as always, with consummate skill. It was shrewd enough to ensure that its voice was heard, but international furore was kept at a level that would not lead to anything like the severing of diplomatic relations or trade sanctions. Neither Singapore nor the United States appeared to have to compromise their respective positions.

Perhaps the whole affair is nothing more than a game of diplomatic finesse, a ruffle in the cocktail circuits of the two capitals. Still the assault on the United States and the West is a bit of an old story in Singapore; it has even become an element of the 'national culture'.

From the 'anti-yellow culture' campaign of the 1960s when long hair and unwashed hippies were seen as American decadence incarnate, to the quarrels with the *Far East Economic Review* and *Asian Wall Street Journal* for 'interfering with domestic politics through biased reporting', the assault on the West and the United States is very much the stuff of confidence-building for the young nation. For this it has to over-read the hostile intentions of the Western Other just as it has to exaggerate its own vulnerability. Excessiveness and over-sensitivity are the rule. If the Fay caning took on the ghostly qualities of a Balinese shadow-play, it is because it is a sign, a symptom of Singapore's dogged, Asian postcolonial identity as it attempts to find its place in the world. The anti-West denouncement is the gesture of a young, dynamic Asian nation, making us different from them, East from West. Yet the denouncement remains something of a feverish imagination, and what happens, we may ask, when the values of the West are already our own?

German sociologist Georg Simmel has the idea of the stranger as someone 'who comes today and stays tomorrow'.[29] When the Western Other is continuously in our midst, he is no longer our enemy; he becomes the classic stranger as Simmel defines it. Enemies are essential, in fact useful, because they help to say something about what we – ourselves and friends – are not and do not wish to be. Friends exist because there are enemies who help to define the boundaries of friendship. Friends and foes are poles apart. The stranger however makes us see the illusion of this symmetry because he is neither friend nor enemy or he can be both.[30] By breaking the neat divisions of friends against enemies, the stranger threatens the certainty of knowing and thus of the social order itself.

Modernity and globalization, when we think of it, are very much about reconfiguring the foundations that have normally defined friends and enemies. The end of the Cold War considerably speeds up the process. To an Asian state like Singapore – and the former Soviet Union for different reasons – the West can no longer give comfort as the Other. The West is no longer the enemy that helps to draw the boundaries of what we are. Simmel's idea of the stranger sums up the ambivalence of Singapore's feeling towards the richest and most powerful nation in the world. What is the nature of the passion that has, since the days of the 'anti-yellow culture' campaign in the 1960s, moved the understanding of the United States as a threat to Singapore's Asian cultural identity, a slayer of its moral centre? The answer is that the United States has become the classic stranger who is at once both friend and enemy, and neither. As the stranger, the

United States/the West is threatening because it takes away an Asian subject's ability to respond since he is never absolutely sure what it is. The reaction, logically enough, is to try to give it shape by resolutely placing the stranger on the side of moral corruption and social laxity: all the things that an Asian subject is not. As a discursive move, the ardent anti-Western posture transforms the West as a *stranger* (back) to an *enemy*.

In this sense the struggle over the right and wrong of the Fay caning is inevitably about the struggle of meaning. For the Singapore State, the West has to be taken to account because of its rapacious transnational movements and the cultural effects they bring. By defining the West as unambiguously the Evil Other, the move aims to create immunity from its damaging effects. The undertaking is disingenuous at best. It is also arguably futile when Singapore is so closely tied, economically and strategically and culturally (by way of the media and popular consumption), with the United States. Perhaps that is why, for all the virulent protests on both sides, businesses were decidedly 'as usual'. Taking a historical view, one might say that the argument over the Fay affair is consistent with Singapore's overall cultural ambitions. Right from the early years, when Singapore eagerly sought foreign investment and to develop a capitalist economy, it foresaw a different cultural destiny for itself. In the 1960s, when it was busily changing the rules of the labour movement and the free press, Singapore had famously worked towards creating its own brand of 'democratic socialism' – the so-called 'socialism that works'. Later, at the height of its economic growth in the 1980s, it introduced Confucian and religious education in the schools and 'Asian Values' as the nation's cultural underpinnings, and panacea for the ills of Westernization. An anti-West stance loomed large in all these enterprises. And presently the anti-West stance is the more aggressive when economic and diplomatic ties with the West are the more intimate. In the Fay caning the driving passion against the United States was largely an affair of *culture*.

To say that the Fay affair is largely one of culture is to point to the fact that the United States matters hugely, and yet does not matter, in the Singapore nation's sense of itself. Like many East Asian societies, Singapore has eagerly pursued capitalist development while putting up barriers against its alienating, cultural effects. Like 'secular Islam' in Dr Mahathir's vision for Malaysia, or Deng Xiaoping's 'capitalism with a Chinese face' for China, Singapore's ambition is to harvest the fruits of capitalism while keenly guarding the presumed Asian cultural and moral sensibilities. Confucian/Asian Values and idea of the family and community are to sit cheek by jowl with lifestyle consumption, Western education, foreign travelling and other things of a cosmopolitan taste: this is the cultural project of the New Asia. And the project sits on a knife's edge, as rising material prosperity brings not only better living standards, but also dangers of cultural and communal breakdown, so the argument goes. No wonder that Asian Values come across as an afterthought, a contemplation without moral curiosity – on the contrasting fates of the East and West.

When the West is figured as Evil Other, we still escape the vacuous questions: are people in the West indeed so loveless, so fond of pleasure and without regard to work and the family, and how do Western values – individualism, materialism and so on – threaten Asian cultural and moral endowment wherever that comes from? The questions are impossible to answer except with a great deal of ideological self-righteousness. At the end, the casting of the West as Evil Other is an attempt to give shape to the dream-like qualities of East and West, qualities that hold true for neither. What is ultimately at stake however is the certitude of meaning as the Singapore State imagines it: meaning about the West, of what constitutes the Asian subject and, of course, of judicial administration of pain itself.

Pain is pain and nothing but...

In the prison room where Fay received his due, the meaning of the punishment is, as they say, a contested terrain. Fay's parents, President Clinton, US television talk-show hosts, human rights activists, government officials and Singaporeans all saw the punishment differently and tried to extract different significance from it. For the Singapore State, those who tried to raise the issues of torture and human rights clouded the issue of what is really a simple matter of the deterring effect of caning. The situation was clearly getting out of hand when a mode of punishment routinely carried out within Singapore's prison walls became a subject of international debate. Singapore's responses were to make a stand and to state its position with excessive passion. The Fay caning must be brought sensibly back to what it originally meant and was intended for. For the State, the point had to be made again and again: caning is a just and legal punishment, and an effective prevention against crimes. There can be no other meaning; Singapore's peace and order speak for themselves.

Hence we can be impressed with the unwavering finality of Lee Kuan Yew's words: '[Caning] is not painless. It does what it is supposed to do, to remind the wrongdoer that he should never do it again.'[31] Similarly, one is not surprised that, when the defence raised as plea Fay's condition of 'attention deficiency disorder' (ADD) which constantly drove him to reckless attention-seeking, it became a subject of derision in the government-friendly media. Pain narrows down the significance that people can read into the mode of punishment: this is what Lee shrewdly recognizes. Even the foreign critics who disagree with Singapore's position will understand body pain as something they do not want for themselves. If they were to put themselves at the end of the cane, they too would feel its excruciating effects. Except for a few masochists, people's instinct to avoid pain of all kinds lends instant logical sense to Lee's argument. Since caning has no other meaning but pain-giving, and since pain deters crimes, the punishment is trimmed of any ambiguities. In the process it cuts liberal 'do-gooders' down to size just as it puts to shame the Western 'social work approach' to crime and punishment.

But the singular meaning of pain is most pertinent for the person who receives it. Meaning is made on his body. On the Western body are circulated wishes and desires of the Asian subject. The administering of pain on a Western body shows up the contrasting sentiments – the Asian state's determination to punish harshly over the West's lesser willingness to do so, the Asian state's keenness to protect communal well-being over the West's readiness to let wrong-doers go. So the Singapore State may well be right: its tough sentencing is designed to extract the utmost 'pain' from the offenders who deserve it. However the State cannot do this without carving on the body, so to speak, all the rich significance of pain-giving punishment.

Imagine, if you will, what went on in the gloomy solemnity of the prison room where Fay received the caning. The first effect is isolation; there is no one to hear the scream (he actually only whimpered, we are told) except himself, the caner and the prison officials in attendance. Isolation ensures that no moral sympathy can come from a viewing public, and Fay suffers alone. Not only is it the mode of punishment, but physical pain even more mercilessly isolates as it inscribes a sensuous experience that only one person can feel. For that carnal experience cannot be shared with another, and it is harder still to communicate by words ('I am in pain; do you know how I feel?'). In Elaine Scarry's majestic words:

> When one hears about another person's physical pain, the events happening within the interior of that person's body may seem to have the remote character of some deep subterranean fact, belonging to an invisible geography that, however portentous, has

no reality because it has not yet manifested itself on the visible surface of the earth. . . .

Vaguely alarming yet unreal, laden with consequences yet evaporating before the mind because not available to sensory confirmation, unseeable classes of objects such as subterranean plates, Seyfert galaxies, and the pain occurring in other people's bodies flicker before the mind, then disappear.[32]

Elaine Scarry is describing the effects of torture. As pain isolates (because only I *feel* it), the effect is very much enhanced by the very conditions of the interrogation where torture is the main instrument. Brought to the prison by faceless men in the dark of night – one thinks of the films of French director Jean-Pierre Melville on the French Resistance – the prisoner is made to feel the uncertainty of his position; he is never sure if he is to be immediately executed or face prolonged torture or left to die in prison. In the prison cell the normal rules of time and expectation and hope are radically changed and perhaps destroyed. Pain inflicted by torture destroys language, and the social universe that gives human conduct its normal, life-sustaining import:

Whatever pain achieves, it achieves in part through its unshareability, and it ensures this unshareability in part through its resistance to language. . . . Physical pain does not simply resist language but actively destroys it, bringing about an immediate reversion in a state anterior to language, to the sound of cries a human being makes before language is learned.[33]

In this sense, pain may be said to take away from the prisoner the freedom to 'interpret' and re-create for him what is of significance. To the body in pain is tied, pathetically and tragically, the loss of personal meaning and the end of the normative world the sufferer once knew. Intense pain is 'world-destroying'.[34]

All this is the language of torture and illegal imprisonment; the Fay case is far less dramatic; nonetheless pain would have similarly torn down his world and what he represented. And generally for the law, this demolition job is undeniably a positive act that protects society from the likes of Fay. When caning destroys, so we reason, it destroys the indolence of an American teenager spoiled by the good life and parental indulgence. After that one may even glimpse at his new self when he reportedly took his punishment 'like a man' and thanked the prison officer. The end of the world of the socially recalcitrant is the beginning of the new self when the harsh punishment turns him around. Strapped to the trestle, with his buttocks naked to receive the lashings, Fay the prisoner is made a passive object to receive his rightful punishment. There is no place for intellectual sophistry and soft pleading in the ritual formality and the efficient extraction of pain. Judicial caning is painful, and pain is nothing but pain: this is the different, more effective Asian approach to judicial punishment.

Conclusion

French psychoanalyst Jacques Lacan has spoken of human desire as driven by the feeling of 'insufficiency'. In the pursue of all the things it 'wants', desire needs to compensate for this 'lack' that lies secretly behinds all the ambitions for fulfilment. It is not too much to speak of the 'internal thrust' of the Singapore State as something that arises from its 'insufficiency' in this sense. And what is this but profound dependence on the West, a dependence that the desire is hard put to acknowledge? The caning of Michael Fay, to render it in the language of Lacan, stages the drama of the State's imaginary movement from 'insufficiency to anticipation', from the singularity of its dependence to the fantasy of its wholeness.[35] As caning is being made to carry rich meanings – as deterrent, as reflection of the State's determination to punish whatever the international criticism, and so on – the State is caught in the seductive lure of its own imagining. The succession of Singapore from a postcolonial state to a First World nation is 'the succession of fantasies that extends from fragmented body-image to a form of its totality', to use Lacan's words.[36] Singapore's nagging suspicion that it is less than what it proclaims, we suspect, is also the Singapore Story – quite apart from the social and economic success. The vulnerability of Singapore may be a social fact drawn from the chaotic, bloody history of 'national struggle'; it is also a construct, a fiction. As with other incidents I describe, the State's responses to the Fay caning smacked of excessive, over-compensatory moves. It is not just that Fay and the likes of him have to be punished and punished painfully, but the meaning of it too has to be managed and unambiguously put to the world. With so much entailed, caning calms the panic of the desiring subject that finds itself in yet another episode of 'national crisis'. The determination to punish, to extract pain from a citizen of the most powerful nation in the world, reveals if only in the psychological realm the truth of Singapore's unwavering confidence and lack of moral self-doubt. In Lacanian terms, the caning of Fay is an opportunity, a pretext, to narrate the Singapore Story: how a Third

World nation has transformed itself by tough policies and a blend of Asian culture and modern capitalist ways. As one would expect, there will always be another occasion for the story to be told. In 2004, Amnesty International released a report *Singapore: The Death Penalty: A Hidden Toll of Executions*.[37] The report gave Singapore the reputation of having the highest execution rate per capita in the world. From 1991 to October 2003, 408 people were hanged, mostly for drug trafficking and murder charges. Surveys conducted in 2003 showed that 'Singaporeans were largely in favour of the penalty'.[38] The government's response was reportedly this:

> Singapore is quite unapologetic about its world beating record. It says its liberal use of the death penalty discourages drug use and violent crime. 'By protecting Singaporeans from drugs, we are protecting human rights,' MP Inderjit Singh said in response to the Amnesty report. 'The rule-breakers have to be dealt with. It's the same in any country.'[39]

5 Oral sex, natural sex and national enjoyment

[A]s far as 'pleasure' is concerned, it may readily be admitted that it is materialist; ... Pleasure is finally the consent of life in the body, the reconciliation – momentary as it may be – with the necessity of physical existence in a physical world.

Fredric Jameson, 'Pleasure: A Political Issue'

The Sex Slave Case 1995

In April 1995, after a 68-day trial, the court found a 24-year-old man, Tan Kuan Meng, guilty of 11 charges of rape, attempted extortion and oral sex, and sentenced him to 13 years' imprisonment and ten strokes of caning. In a country where *Playboy* and *Penthouse* magazines are banned, where net access to pornographic sites is restricted, the case enjoyed a certain public notoriety. For a few weeks newspaper readers were given sensational accounts of sexual misdemeanours committed under lurid circumstances. Tan, a dispatch rider – whose photo in the paper shows a man with a flabby face and sly, sheepish eyes – had taken the victim, a 22-year-old woman clerk, to various hotels and subjected her to his demands for 'sexual perversions, sex and money against her will'. The offences took place over a period of five months from April to September 1993. The court judgment described Tan in no uncertain terms as a 'manipulator' working his callous charm on a 'timid and somewhat weak-minded' woman. 'She was unwilling to engage in any sex with you or to part with her hard-earned saving. You drove her to the end of her tether', the presiding Justice Lai told the accused. 'Timid, timorous, and all the more even weak-minded girls have to be protected from people like you', he said.

In all, Tan extorted a total of $13,150 from his victim for which he received five years and two strokes of caning. The more serious sentencing was given for two counts of rape – ten years each and six strokes of the cane. For the five charges of oral sex offences, a total of 17 years of imprisonment was given to Tan, who was ordered to serve some of the sentences consecutively.[1]

In Singapore, judicial ruling on oral sex relies on the Indian Penal Code – a legacy of the early years when from 1834 to 1867 the colony was administered by the East India Company from Calcutta. Section 377 of the Penal Code covers all acts of 'sexual intercourse against the order of nature'. In quaint Victorian wording, it states:

> Whosoever voluntarily has carnal intercourse against the order of nature with any man, woman or animal, shall be punished with imprisonment for life, or with imprisonment for a term which may extend to 10 years, and shall be liable for a fine.

This section clearly covers anal and oral sex between man and man, man and woman, and acts of bestiality. The Indian Penal Code also singles out more generally man-and-man homosexual offences of various sorts. Section 377(a) is designed for such a purpose:

> Any male person who, in public or private, commits or abets the commission of or procures the commission by any male person of, any act of gross indecency with another male person, shall be punished with imprisonment for a term which may extend to 2 years.[2]

It is notable that, for offences being charged under these sections, consent is not a defence, and neither is the fact that the act is conducted in private. The 'gross indecency' of Section 377(a), as a local legal expert explained to me, is a broad term which covers, from the experiences of past charges, 'mutual masturbation, indecent contact at the groin, and lewd proposals even when they do not lead to anal penetrative act'. For lighter offences, however, the judge has the choice of turning to Section 354 dealing with 'outrage of modesty' by assault or use of criminal force. The punishment prescribed under this section is lighter, a maximum two years' imprisonment, with a fine or with caning, or with both.

Tan Kuan Meng was charged under the more severe Section 377 which, we recall, could put away an offender for life. From the newspaper reports, Tan came across as a distinctively unsavoury character. Subjugating a 'weak-minded girl' to his lust and financial demands says enough of his lack of moral scruples. He may have deserved what he got; the charges against him inevitably affect all similar acts of 'unnatural sex' as well. To the gay community especially, the carrying over of a nineteenth-century law of colonial India to the present day, and the intervention in private, sexual enjoyment even among consenting adults, comes across as excessive if not slightly comical. However, there can be no doubt that, for the State and legal circles, and not least the defendant Tan, this is a serious affair indeed.

Oral sex and natural pleasure

Professor Koh Kheng Lian of the National University of Singapore is an authority on Singapore's criminal law. In 1996 following the Tan sentencing, she published in a law journal an article that directly comments on the Sex Slave Case. There she argues that oral sex does not come under 'sex against the order of nature' as intended by Section 377. After examining Indian cases, she concludes that '[t]he type of unnatural offences clearly contemplated under Section 377 are buggery or sodomy by a man per anum with a man or woman, or intercourse per anum or per vaginam by a man or a woman with an animal'.[3] It is these offences, rather than fellatio performed by a woman on a man, that are 'unnatural sex' proper according to the original spirit of the Indian Penal Code. Then there is the even more intractable issue of 'consent'. Professor Koh again clarifies: 'Two ingredients are required under section 377: (i) carnal intercourse and (ii) intercourse against the order of nature. "Carnal" means lustful. Consent is irrelevant – this means that both partners are liable, whether or not there is consent.'[4] Quite simply, since consent is irrelevant, Section 377 makes criminals of heterosexual couples who include oral and anal sex in their repertoire of enjoyment in the bedroom.

Professor Koh is making an argument about the inconsistency of the Penal Code when applied in more liberated, modern times. Looking at the law as it stands, it seems to defy common sense. Not only the legal circles, but the presiding judge of the Sex Slave Case too was sorely taxed, confessing that it had been 'a difficult case'. 'I must tell you that it caused me many sleepless nights thinking of the evidence again and again', Justice Lai Kew Chau said. However, until the parliament repealed it, Section 377 had to be upheld. Nevertheless Justice Lai had to make a judgment that would not haul husbands and wives to jail. As though he had the enjoyment of married couples in mind, he made a ruling that has become something of a classic, as the *Straits Times* reports:

Husband and wife who indulge in oral sex can rest assured that they are not committing an offence....

The judge said husbands and wives could indulge in oral sex with impunity. Not every act of fellatio, he said, was a criminal offence if it was a prelude to natural sex between two consenting adults, and not performed as a substitute for natural or consensual sex.

In Justice Lai's own dignified words,

In my view, an act of fellatio which is performed between a man and a woman as a substitute for and not as a prelude to and enhancement for natural sex between them is carnal intercourse against the order of nature and punishable under Section 377 of the Penal Code.[5]

Sexual pleasure and its use

One needs a lurid imagination to ponder on the implications of all this. The comedy *sans savoir* is as banal as it is of grave consequence. Given that oral sex is legal only as 'a prelude to and enhancement of natural sex', what of premature ejaculation? What is the position of law when the heterosexual couple finds fellatio so intensely pleasurable and satisfying that they do not go on to the subsequent penile–vaginal consummation? Further still, would it be a prosecutable crime when, in a divorce proceeding, it is revealed that the married couple had engaged in fellatio – or other unnatural carnal pleasures – in their happier times?

While laypersons puzzled over the judgment and enjoyed its locker-room humour, legal professionals would seek to clarify the many ambiguities of the Sex Slave Case. A Singaporean lawyer explained to me in an email:

For both Section 377 and 377(a), there is no dispute about consent, and it is immaterial if the defendants are adults. Certainly the Sex Slave Case has established that oral sex, or any form of unnatural sex, is a prosecutable crime when consenting adults do it in the privacy of their own bedroom. So you can't use any of these, mutual consent, legal age and privacy, for defence. The law regarding 'sex against the order of nature' is archaic in the modern times. When you legislate what consenting adults do in private and when the act does not involve injury to themselves or to a third party, then you have a very difficult situation indeed.

Against this sort of contention, Justice Lai's ruling really does not provide firm guidance. As the *Straits Times* reports,

> When asked about oral sex between two consenting adults who did not proceed to have natural sex, the judge said: 'What is wrong with that?'
>
> Counsel wondered if the court was saying that so long as it was between two consenting adults, it was not an offence. The judge told him 'Let us cross the bridge when we come to it.'[6]

The retort sounds like an admission – that given the law in the present form ambiguity will be there. For what Justice Lai has been asked to settle, we might say, is primarily philosophic questions – quite apart from the interpretation of law. These are about what makes oral sex 'unnatural', the origin of its social and moral harm, and the quality of its self-affirming pleasure. And they would instigate questions about the opposite: what makes the 'penile–vaginal' sexual act natural and thus socially useful?

As always with Singapore, it is not going too far wrong to examine this kind of query with a practical eye, to seek an answer from the view of the State's policy needs. In regard to 'sex against the order of nature', the spirit of the Indian Penal Code is concerned with intercourse that does not lead to conception, as another case that came before the court confirmed, as we shall see. Since nature did not intend the mouth to be of sexual and reproductive use, so the state prosecution argued there, fellatio is 'unnatural'. Reproduction – the proper function of sex as nature intends it – is the crux of the matter, one closely related to Singapore's celebrated population policy.

Madame La Terre, the better half of Monsieur Capital, does not smile kindly on Singapore. A small island at the south of the Malay Peninsula, it has no natural resources, and food, water, oil and raw materials for industries have to be imported. Partly for this reason, the small island nation is concerned, to the point of obsession, with keeping up the level – and quality – of its workforce. 'Singapore has no natural resources except human resources', the official refrain goes, without a touch of irony. Speaking of human beings as equivalent to so many tonnes of minerals or kilolitres of water is, in the bureaucratic view, merely expressing the island's special demographic needs. In recent years, Singapore has increasingly been looking at Hong Kong – another successful Chinese port city in East Asia with a similar industrial base – as a model of future development. City planners hope to create a vibrant city of high population density to stimulate competition, and to lower the average costs of city infrastructure. Crowded urban living, in their vision, does not translate into social malaise and services breakdown of the sort one encounters in Calcutta or Mexico City, but a jungle of healthy 'Darwinian struggle' in which keener and smarter people survive and take their rightful places in the higher echelons of society.

But even before it saw itself as a 'hot-house' global city, Singapore had famously taken to devising a population policy to serve its special 'human resources' needs. In the 1983 National Day rally, the then Prime Minister Lee Kuan Yew told the people of his worries about the declining birth rate of young graduate mothers:

> This was a problem, Lee reasoned, because graduate mothers produced genetically superior offspring.... Eighty percent of a child's intelligence, Lee explained, citing certain studies in genetics and sociobiology, was predetermined by nature, while nurture accounted for the remaining twenty percent. With a few generations, the quality of Singapore's population would measurably be swamped by a seething, proliferating mass of the unintelligent, untalented, and genetically inferior: industry would suffer, technology deteriorates, leadership disappears, and Singapore loses its competitive edge in the world.[7]

Human intelligence as genetically inherited, as largely endowed by birth, is controversial stuff and it immediately drew criticism. For the progressive circles, the lament of a low birth rate among graduate women, and that women should find a role in the nation's economic needs, is at once to stack the disadvantages of class and gender against women. And since graduate professional women are mostly Chinese, race too became an issue. The infertility of graduate women – as carriers of intelligence – was deemed irresponsible and as undermining all that the State was trying to achieve:

> The indictment of women, then – working class and professional, Malay, Indian, and Chinese – inscribes a tacit recognition that feminine reproductive sexuality refuses, and in refusing, undermines the fantasy of the fertile body machine, a conveniently operable somatic device: thus also undoing, by extension, that other fantasised economy, society as an equally operable contraption. Indeed, the disapproval, simultaneously, of an overly productive and a non (re)productive feminine sexuality registers a suspicion of that sexuality as noneconomic, driven by pleasure:

sexuality for its own sake, unproductive of babies, or babies for their own sake, unproductive of social and economic efficiency.[8]

With so much attention given to female productivity, the graduate mother debate gives full rein to the State idea that female sexuality has to tie in with the wider demographic needs of society. Women were given this national responsibility, and their sexual enjoyment has to curbed; this is what irked the critics.

Private ecstasy

As its turns out, the 'reproductive policy' eventually put in place is a good deal more varied and politically astute than Lee's words suggested. His idea of human intelligence as a matter of biology, and the burden he placed on women, naturally fired the passion of the critics. And there is a formal elegance when they took Lee's words as they were, seeing them as promoting favouritism towards one group – and disparaging unfairness towards another – in terms of class, gender and race. To take Lee's words too literally is to uphold his crass imagination of sexuality's use. This eventually led the critics to a bit of a dead-end. Since human fertility is about sexuality, and sexuality is about private ecstasy, then the State would have to deal with the intense, lustful pleasure of a man and a woman as well. That, I am sure, is the more primary issue. Perhaps it is pleasure, not so much fertility, that invites the State's censure of unproductive sex.

In any case, what is clear is that the population policies are periodically fine-tuned and have their more unpleasant features removed. For instance, at the beginning monetary rewards – in the form of superannuation deposits – were given to mothers of low-income families who accepted sterilization. After wide controversy, when people spoke of such rewards as 'blood money' in exchange for forced infertility, the government eventually abandoned the scheme. The population policy is not all class and gender prejudices. The unlikely success story has been the Social Development Unit (SDU), an initiative of the Department of Home Affairs. The SDU is a match-making agency to serve unmarried graduate men and women. Run by a woman with a doctorate in psychology, it organizes state-subsidized afternoon tea dances, etiquette and grooming lessons, and a 'mystery cruise' in the moonlit tropical night around the nearby islands to foster romantic liaisons of professional men and women often too busy even for such things. An internet discussion web site and 'profile compatibility' testing have now been added to the services. In 2002, the SDU announced that

4,000 members had 'tied the knot' and nearly 30,000 have done so since 1984.[9] I have visited one such couple; Richard and Jenny Lim speak to me on the balcony of their Housing Development Board (HDB) executive flat of their meeting at the computer training class where they found themselves sitting next to each other, and how sharing the keyboard led to friendship that finally blossomed into matrimony. Now they are proud parents of a six-month-old baby. As the maid prepares the child for church, the domestic normality seems light years away from the artifice of State match-making, and the population policy with its genetic racialism that drives the passion of many critics.

This is, however, not to make light of the ethical problem – and ideological intent – of something like the SDU. Nevertheless once pruned of its more controversial aspects, the State policy can still be something many seeking to be married find appealing; they may even enjoy the midnight cruises and afternoon tea dances. When we talk of the State regulation of 'unproductive sex', it needs little reminding that Singapore is no sexual desert. Pleasure of the 'oldest profession in the world' is available in the dark alleys of Desker Road near 'Little India' where foreign labourers from South Asia gather, and in establishments in Geyland, favourite haunt of local men. Gay men and women have their clubs and bath-houses where lustful encounters are consummated with the help of an ecstasy pill or two available even in Singapore.[10] The fact is that, for practical reasons – foreign labourers are admitted to Singapore without their spouses – and in the current liberal climate, the State quite often turns a blind eye to what is happening in these places. The sex industry is not so much hidden as heavily regulated; it is being confined to certain areas, making it easier for the Health Department to carry out health checks on sex workers and to ensure they do not offend the public by openly soliciting customers.

In the light of these realities, the State's anxiety about the low birth rate of graduate mothers, and the ruling against 'unnatural sex' seem hard put to be aligned with the practical needs of population policy. Turning to the Sex Slave Case, what then is the logic of the prohibition of 'sex against the order of nature'? Of course, when we make too much of the instrumental reason of the State, then the suppression of 'unnatural sex' is simply a deterrent. Similarly, when we believe that the sentencing of the Sex Slave Case is to repress a perverse sexual practice that is 'wasteful of seed', then the State has to, logically, legislate what is and what is not allowed in the bedroom. These are difficult conclusions to make. Does the State really imagine that the harsh punishment will turn people away from the carnal pleasure that the law does not formally permit? People, in a moment of lustful intensity – in normal or abnormal conjugation – are not likely to be mindful that oral sex is only legal when it is foreplay, when it is a prelude to vaginal-penetrative sex.

The first volume of Michel Foucault's *History of Sexuality* is the slimmest and probably his most accessible work. In nineteenth-century Europe, he suggests, sex was marked by prudery, repression and obsession with bourgeois respectability. The same period also saw a great deal of 'talk' of sex, of its pleasure and corruption, moral danger, and the need to rein in its excesses for the health of individual life and for reproduction. Seemingly dominated still by Victorian values, we are preoccupied with sex today. However, the argument about social and legal repressions really does not sit well with the busy, organized and elaborate 'talks' – what Foucault calls discourse – of sex, some about its moral danger, some instigating the hypocrisy in our attitude. 'If sex is repressed,' Foucault writes, 'that is, condemned to prohibition, and silence, then the mere fact that one is speaking about it has the appearance of a deliberate transgression.'[11]

For Foucault 'the repressive hypothesis' in understanding Victorian sexuality is not so much mistaken as asking the wrong kind of questions. For discourse itself *offends* the regime of sexual prohibition. Such is the power of discourse that we cannot talk about sex without being aware of the effect of what is being said, without being 'incited' by the nature of the pleasure we imagine sex to offer. The point is not whether sex is denied, silenced or repressed, but the way it is being spoken about and the people and institutions that do such talking. In Foucault's felicitous prose,

> my concern will be to locate the forms of power, the channels it takes, and the discourses it permeates in order to reach the most tenuous and individual modes of behaviour, the paths that give it access to the rare or scarcely perceivable forms of desire, how it penetrates and controls everyday pleasures – all this entailing effects that may be those of refusal, blockage, and invalidation, but also incitement and intensification: in short, the 'polymorphous techniques of power.'[12]

This justifiably famous passage reminds us that sexuality – and our understanding of it – is intimately connected with power and knowledge of all sorts. Power is more than domination and physical coercion, but also the enabling stuff that the weak also eagerly seek for themselves. Sex is a similarly complicated affair. Turning Foucault's insight to the Sex Slave Case, we say that oral sex has to be managed not because it is a 'stubborn drive' naturally getting out of control for lesser men and, worse, because it undermines demographic planning. Oral sex has to be tamed and controlled because so much is a part of it: it is 'a dense transfer point' – to use Foucault's spatial metaphor – through which flow diffuse and knotted ideas that centrally matter to the State. To continue in the same vein, fellatio is the traffic light on a busy highway that guides and controls the frantic passing of thoughts and desires. What renders oral sex a subject of obsessive attention is because it is a key node on a map, Foucault might say, and across it moves an unruly company of travellers: order and social recalcitrance; the contrasting ideas of 'unnatural' and healthy, reproductive sex; private pleasure and collective responsibility; State power, its impotence and mendacious intent; and, not least, nature and what human understanding would have it.

It is a picture of fluid entanglement indeed. For the State, 'sex against the order of nature' is important not because those in power are prudish, or because they badly need to tie sex to reproduction. Rather it is because fellatio and other forms of unnatural sex 'incite' other ideas and forces, bringing them to the fore to influence what people do and say and wish. To see suppression of oral sex from the point of view of population planning, as the critics do, is to give in to the fetishism of instrumental reason so much a part of the State discourse. Instead, 'oral sex' defies the practical, economic needs of the State, as Foucault would say. If that is the case, we have to ask: what anxious discernments flow through, and are luxuriantly fed by, the excessive pleasure of oral sex?

'The fate of pleasure'

In the busy traffic with other forces and ideas, 'unnatural sex' is neither straightforwardly repugnant to social morality nor self-affirmingly enjoyable. The prosecution of oral sex, when the State takes upon itself to do so, forewarns the public of the moral danger and social and biological wastefulness, but the carnal pleasure is obviously more than that.

Speaking of pleasure generally, the history of Western thinking has swung between what American literary critic Lionel Trilling calls 'two separate moral ambiences and two very different degrees of intensity'.[13] Western philosophy has historically tended to regard pleasure with ambivalence. On the one hand, there is the puritanical distrust of pleasure as lustful, voluptuous indulgence without restraint. The other, by way of the Romantic vision, views pleasure as positive delight of cultural delicacy and natural sensibility. In the Romantic vision, pleasure, aided by the arts and nature's benevolent influences, helps the making of a virtuous, contemplative self. For its negative aspects, the word 'pleasure' has a history of unfavourable meanings, as 'indulgence of the appetites and sexual gratification'.[14] When it is linked to the Latin word equivalent *voluptas*, pleasure and the people who seek it are put on a lowly moral scale; those who enjoy voluptuous enjoyment of various forms do not have in mind the quiet pleasure of reading by the fire, contemplative walks in the woods or, we should add, duties to society.

These are the twin fates of pleasure, one of moral and social sensibility, the other bodily indulgence. For convenience we call one enjoyment, the other appetite. Appetite is a kind of degraded passion, a self-regarding sensuality 'taken [by the subject] as an end in its own right'.[15] Clearly, in the Sex Slave Case, it is appetite that is the worrying thing. And appetite – the lustful, private pleasure – has to be regulated because it is restlessly expansive. If pleasure can be enjoyed in excess, then as appetite it tends to undermine other values, other ideas of individual obligations and duties of the State. As it nestles in the network of these discernments and ideas, 'sex against the order of nature' opens up and then redraws the boundaries of the rules of power and moral normality. What is at stake is not, literally, the repression of the immoral and unproductive *appetite*, but the need to regulate the flow of different and dangerously contrasting meanings of pleasure.

And here, pleasure and appetite are no longer the abstract stuff of philosophers but practicalities of the everyday; they are also something related to the State design. Among other things, what makes the regulation of appetite so important to the State is because it has since independence made 'national enjoyment' a central part of its policy. As I have shown elsewhere in the book, the delivery of health services and education, peace and order, employment and economic growth gives powerful legitimacy to the PAP rule and its tough measures. These benefits work on the 'heart and mind', recruiting collusive support from Singaporeans whose gratitude returns the government to the parliament with a huge majority term after term. But the rub is that 'national enjoyment', as Trilling reminds us of the 'fate of pleasure', tends to spin out of its orbit of moderation and reasonable expectations. When that happens it is hard for the people to remember that the good life, the employment and overseas holidays and the abundance at the shopping malls are the works of the PAP government.

Harder still is for it to impress upon the people that the enjoyment of these goods is not a 'right', but an 'entitlement' that goes with the duties and obligations of citizenship. In Trilling's phrasing, the State must prevent enjoyment from erupting into appetite. When that happens, it creates spurious problems for 'national enjoyment' meant to be translated into electoral victory. If bringing oral sex offenders to court sounds like a last-ditch, desperate act, it is because cases like this are a sign of social danger seemingly lurking behind the delight of the bedroom and, in the wider sense, a worrying vista of how 'national enjoyment' can go wrong. In this light, the laxity of 'Let us cross the bridge when we come to it' in the judgment of the Sex Slave Case looks increasingly like an admission of the State project's potential weakness. With so much at stake, such open-endedness of judgment will not do; something more rigorously definitive will be needed.

Nature and the state ruling

In September 1996, a year following the Sex Slave Case, a man appeared before the court charged with having 'unnatural sex' with a 19-year-old woman. The offender, Kwan Kwong Weng, aged 47, was accused of making the girl perform oral sex on him twice in a hotel room in March of the previous year. Kwan and the girl worked in the same management firm where he was a technician and she a temporary clerk. Apparently she had had oral sex with her boyfriend, and Kwan then made her believe that the swallowing of semen had admitted poison in her body. Kwan convinced the young woman that he had the medical skill to extract the poison from her; and the therapy was for her to fellate him. They went to a hotel where the offence took place. Later the girl realized that she had been deceived and reported Kwan to the police. He was charged under Section 377 of the Penal Code which deals with, we recall, 'sex against the order of nature', an offence carrying a sentence from ten years to life imprisonment.

In the judgment, the Judicial Commissioner had to wrestle with the principle established by the Sex Slave Case. Unlike the previous offender, Kwan did not intimidate the girl or subject her to threat and violence. Without threat and violence, the offence could not be rape. The circumstance makes the Sex Slave Case a poor guide; and one detects a note of exasperation in the Commissioner's words:

> When does an act of fellatio or oral sex become a lustful substitute for and not a prelude to an enhancement of natural sex? What if one consenting party subsequently changes his or her mind about

proceeding to have sexual intercourse? At what point does an act of fellatio or oral sex stop being a prelude to or enhancement of natural sex? How long must the interval or interlude last between the prelude and the act of sexual intercourse?

In raising the haziness of the 1995 case, the Commissioner points to the only mitigation with which oral sex can escape prosecution. 'Prelude to penetrative penile–vaginal sex' turns fellatio around, giving it nature's stamp of approval because it is 'used' to aid the consummation of normal, heterosexual intercourse. The ideas of the 'natural' and 'unnatural' are very much the issues. The defence had turned to the argument of Professor Koh Kheng Lian of the National University of Singapore who has, as we have noted, commented on the Sex Slave Case. Regarding the principles of the Penal Code, she explains that the Indian court had traditionally 'proceed[ed] on the premise that [oral sex] is unnatural because it falls outside the natural object of intercourse which is that there should be the possibility of conception of a human being'. What renders oral sex 'unnatural' is that it does not lead to conception. This principle she finds both 'narrow and artificial':

> While sexual intercourse leading to conception is in the nature of things, it does not necessarily follow that intercourse which does not lead to conception is unnatural. The most obvious example is where the woman is infertile or birth control measures are used. Does that mean such intercourse is unnatural?[16]

If we were to take 'conception' as the deciding principle, Professor Koh argues, then 'sex against the order of nature' would catch within its net infertile women, and heterosexual couples practising birth control of various forms, including, one suspects, *coitus interruptus*. She then goes on to cast doubt on the idea that fellatio is 'intercourse' at all. In truth, since the act involves the mouth, the technique of pleasure-taking resembles more of a kiss:

> [It] seems a strain on the ordinary use of the word 'intercourse' to describe oral sex as an act of intercourse. As the mouth is being used, it would be more appropriate to classify such an act as a kiss in ordinary language.... If the legislature had intended to include oral sex in section 377, it should define 'intercourse' to include it.[17]

Perhaps swayed by the expert opinion, the Commissioners cleared Kwan of the two charges of 'unnatural sex'. The Commissioners also ruled that oral sex between consenting male and female does not come under Section 377 of the Penal Code.

Kwan's dismissal undid in one stroke the skilfully assembled principle of the Sex Slave Case. Instead of the clumsy 'prelude to natural sex', a more straightforward reasoning of 'consensual enjoyment' between adults now renders fellatio legal. The ruling applies however only to the act between a man and a woman, and not to that of man and man. With woman–man fellatio removed from it, Section 377 is left to more strictly deal with 'sex against the order of nature', including sodomy between man and woman, and between men, as well as oral sex between men. When violence is not involved, when there is mutual consent, the Judicial Commissioner argues, oral sex offences should be more properly charged as 'outrage of modesty' under different sections – Sections 354 and 354A – of the Penal Code. The offender Kwan should be so charged because the girl had apparently agreed to the act without coercion.

For the lawyers I talked to, the Kwan case cleared up the knotty issue of under what circumstances oral sex between a man and a woman could escape prosecution or be an offence. What is left hanging, however, is the crucial question that Professor Koh has raised, one underlining the spirit of the Penal Code: what makes one form of sexual act 'natural' and another 'unnatural'? The question comes about because conception is a poor determinant. Three months later in November 1996, the State prosecutor brought the case before the Court of Appeal; and it was primarily to this question that he turned.

For the Deputy Public Prosecutor, the root of the argument lay in the notion of 'nature'. In his view both the Sex Slave Case and the Kwan case overturn the fundamental spirit of the Indian Penal Code. What these cases say in effect is that reproduction or conception is *not* the defining thing of 'natural sex'; this is wrong. Not surprisingly, he then turned to biology in order to make his case. In a piece with the heading 'DPP argues only Parliament can make oral sex legal', the *Straits Times* reported:

> [The Deputy Public Prosecutor's] view was that the courts have no room to interpret existing law to mean that oral sex is all right, even where there is consent between partners.
>
> The relevant legislation has been in use for 125 years, he said, and only Parliament could change it to state that oral sex is no longer considered an act 'against the order of nature'.[18]

Like Justice Lai earlier, the DPP found himself having to take on the ponderous task of metaphysics. Appealing against Kwan's acquittal, his 78-page submission 'cited numerous dictionaries and legal references'. Casting aside the two oral sex cases, the DPP asked the appeal judges to uphold the definition of 'natural sex' in the Indian Penal Code: that 'the only form of human sexual intercourse that is not against the order of nature is male–female intercourse via the vagina and which can lead to conception'. All other acts of sexual penetration involving the male organ are 'unnatural'. Changing social norms do not alter this principle; he then goes on to a bit of anthropology:

> What is in accordance with the order of nature or against the order of nature does not depend on the changing practices and beliefs of the peoples or tribes at any point of time.
> It is simply dictated by how nature has made the human body and the sexual organs.[19]

What is 'natural' then is simply what exists unambiguously 'in nature', so he argued. Physiological functions of human organs cannot be swayed by changing culture and habits and, we should add, legal arguments. 'Nature' is the final arbitrator for settling the question of what an organ is originally and rightly intended for: the mouth for eating, vagina for sex and reproduction, anus for discharge of faeces, and so on. Once the function as 'nature intends' is defined, any idea about the moral meaning of a bodily act quickly falls into place. In regard to fellatio, no intellectual debate is needed to decide on its 'unnaturalness' and thus criminality. Fellatio is 'unnatural' because nature has not intended the mouth to be an orifice for sexual intercourse:

> The mouth as much as the anus are [*sic*] clearly not made by nature to be sexual organs, unlike the vagina.
> [Fellatio is clearly against the order of nature] because the orifice of the mouth is not naturally or biologically meant for sexual intercourse.
> Its biological function is related to the digestive system and the penetration of the orifice of the mouth by the penis cannot under any circumstances lead to conception or the possibility of conception in the normal biological course of events.[20]

At the end of the hearing, the Court of Appeal overturned the Kwan case and sent it back to the lower court for sentencing.

History of national enjoyment

There is nothing more illustrative of the self-possessive character of the State than this attempt to define the functions of human physiology that even nature finds hard to do. That the mouth can only be of the 'digestive system' speaks of a meek sexual imagination, and to decree that its orifice is not for intercourse is to rule out the practical possibilities *au naturel* among animals and humans. But the Deputy Public Prosecutor has no pretence to philosophy. The solemn determination settles the singular truth about a practice too unruly and lustily enjoyable for the State's liking. Yet it is hard to reckon that the overturn of the Kwan case is meant to be, more purposefully, a deterrent. After all, similar pleasures are available at bath-houses and massage parlours for those who want it. The fact is that prosecution of crimes of 'sex against the order of nature' is a highly selective and an infrequent affair; they do not come before the court like traffic fines or burglary charges. A rare event, it attracts huge public interest when it happens, creating for the time a climate of fear especially for gay men. Accustomed to the government's ways, Singaporeans take the prosecution seriously. Outsiders may point to the poor practical logic, but the local people are likely to see the outcome of the case as a sign of the State's willingness to apply, as it often does, harsh measures when necessary. When the oral sex ruling makes the point, it does so by its tortuously difficult moral and practical reasoning.

If Foucault is right, then bringing unnatural sex offenders before the court is less about prohibition than about managing the criss-crossing of meanings and desires that something like fellatio gloriously brings up. 'Unnatural sex' invites the State's attention, not because it is unproductive or morally repugnant – though some in the government may think that – but because it is richly suggestive of ideas that hugely matter, ideas that coalesce around the notion of pleasure. In our own terms, what drives the ruling against 'unnatural sex' is the attempt to guide pleasure from the shadow of lustful, self-regarding *appetite* to the light of *enjoyment* – that morally considered, socially minded experience of the senses.

Pleasure is an important issue because since self-government in 1957 the Singapore State has made the provisions of social and material life – from jobs and public housing to education and health care – its main responsibility. These provisions are not all conspiracy and ideological intent, but enjoyed by the people with gratitude. Nonetheless, as a part of 'national development', forms of social and personal enjoyment – including those of the bedroom, as we see – are brought, willy-nilly, in line with the wide designs of the State. Some enjoyments are encouraged: of the family, communal and religious festivals, and patriotic fervour; others of a more private and hedonistic nature – pornography, 'unnatural sex' and unfettered personal freedom and intellectual expression – are dampened. This management of pleasure has a history, a genealogy of ideological design and political purpose.

If Singapore is renowned for the restriction of chewing gum sales, or the censoring of *Titanic* because of Kate Winslett's breasts, it is even more justly famous for its social policy.[21] At a time when the welfare state is going through profound crisis in the West, some commentators are turning to Singapore's approach with interest.[22] The difference between the Western practice and Singapore's 'Asian approach' is really over the priorities given to economic growth and social redistribution. In Britain and Australia, Fabian socialism was the root of state welfare built on 'the democratic vision' of 'taming capitalism through redistributive social policy'. The British Labour thinking of the 1940s was very much that 'the worst of economic evil of capitalism could be controlled by Keynesian demand and management together with selective nationalisation to stop monopoly'.[23] This can be achieved by taxation without complete state economic ownership and control. Social policy, while pursuing social equality and welfare delivery, was to address the failures of the market by achieving fair distribution for all.

In the Cold War years of the 1950s, with full employment and economic growth many in the West thought that state ownership was both unnecessary and inefficient. The result was to move towards the 'welfare state' as we now understand it. The aim was social redistribution, and correcting the uneven impact of the market on different groups of people. The state largely put aside the productive role. In simplest terms, the current debate on economic rationalism and neo-liberalism is also about how to bring 'the economic' back into social policy. Of course, one should immediately add, 'the economic' here is no longer about state ownership. Far from it, it is concerned with allowing the free play of the market over and above redistribution. Social welfare is seen not as a right for all by virtue of citizenship. It is perceived, in the case of the unemployed for instance, as an entitlement for those who actively seek to (re)enter the market economy through employment or some small business schemes.

In contrast, Singapore's social policy shows an ingenious marriage of social delivery and economic ownership. While retaining some direct ownership of old, the State enforces economic rationalism with a rigour unmatched by what is practised in the contemporary West.

Receiving the meagre welfare payment is akin to committing a mortal sin at a time when the State can best afford it. In the past PAP Old Guards had spoken exuberantly about Fabian socialism, which they had encountered in the lecture hall of the London School of Economics and other places during their student years in post-war Britain. They were deeply impressed with the social programmes of the Atlee Labour government that won the first post-war general election. As they returned to Singapore, the condition of mass poverty and unemployment, and no less the reason of the PAP's political survival, made the democratic-socialist principles at once urgent and necessary.

As the PAP leaders have since often asserted, Singapore had to put the economy above all else. Economic improvement, employment and industrialization were urgent priorities, and the nation's viability depended on them. A strong economy made social policy 'affordable', and paid for the infrastructural improvements, public health and the education system. The strong economic bent of democratic socialism was also ideologically pertinent for fighting communism, by taking the wind out of the sails of its popular support. The official sentiment of the time was very much that:

> What the masses want to know is whether Socialism can orga-
> nize rapid economic development. They are not very interested
> in democracy, nor will they fear communism, for it is hard to
> impress a hungry man or his poverty-stricken family with ar-
> guments about the loss of personal freedom, when communism
> provides his neighbours with food, shelter, a piece of cloth for the
> body's nakedness, education for his children, some form of social
> security.[24]

In any event, the social policy that has since arisen in Singapore is one that aggressively ties redistribution to the market and the economy. What this means is that the forms of assistance to the poor and the needy, paltry as they are, must not kill incentives for work and economic self-improvement. They are not to encourage people to 'withdraw' from the workforce, thus risking creating as they do in the West a 'welfare-dependent' jobless underclass. The system is a mutation of the democratic socialism of the 1960s. The State assumes some ownership – rather than nationalization – of industries, while it continues to deliver heavily subsidized public housing, health care and education rather than individual welfare payments. Neither 'the welfare state' nor 'the economic state' of post-war British Labour describes the present system, but it is a deft mixture of both.

Excessive enjoyment

This is the genealogy of the 'national enjoyment' in Singapore. Like the social policy from which it has sprung, 'national enjoyment' is at once economically sensible, ideologically astute and designed to deliver benefits to the people. Things of 'national enjoyment' are not only public housing and health care and such like, but also job security and social peace and, not least, rich offerings at supermarkets and shopping malls – the good life, in short, created by an efficient government. The social policy may be about softening the social damage of the market economy; it undoubtedly generates a 'culture of gratitude' among the people. For all that has been said and written about the PAP government, tough polices do not explain its hold on power, but a subtle blending of them with tangible benefits which people recognize and enjoy. In this sense the 'culture of gratitude', while it ensures the legitimacy of the State, also reveals people's complicity in the making of their own political fate.

However, the trouble with 'national enjoyment' is that people are prone to forget what it is supposed to be – a socially considered enjoyment and an impulse that boosts the 'culture of gratitude' so crucial to the State rule. Here lie some of the problems. As the delivery of the good life accrues legitimacy and popular support, it also tends to encourage unreasonable expectation of the State and what it can do. When the State is asked to deliver what it is not prepared or unable to, such a demand comes dangerously close to suggesting a lack of self-reliance and an over-reliance on the State. Thus we often see a certain irony in the official responses. The State is proud of its social and economic achievements, and has built power on them, yet it must not encourage over-dependence on the State. The result is that the State often has to curb people's belief in its near omnipotence to deliver. Yet since the State claims all the credit for what Singapore is today, it can hardly be too dampening of the idea that it can meet all kinds of needs and fix all kinds of problems – from assistance to businesses to fostering creative thinking and remedying social problems in the ethnic communities. For the State to do so is to upset the foundation of its legitimate rule and the root of its 'social-democratic' policy.

What happened in the recent economic conditions is a good example. In 2002 when Singapore faced rising unemployment and a contraction of the economy, the State continuously assured the people that, in

the words of the then Prime Minister Goh Chok Tong, 'it will look
after them like members in a family'.[25] The Singapore economy is
traditionally dependent on export, especially to the United States; it
is closely tied to the health of the global economy. But it will not do
for the State to tell the people that the economic trouble is beyond
its control, that it is due to a decline in the world – especially the
United States' – demand for Singapore's exports of hard disk drives,
petroleum chemicals and other high-technology products. To suggest
that the economic recession is due to 'external factors' may be realistic,
but it does not sit well with a society accustomed to think of the PAP
as responsible for all good things in the country. Putting it simply,
taking all the credit would mean that the State would have to take the
blame as well. So in a time of recession, the State would assure the
public that it was taking tough measures to aid economic recovery.
People should be patient, and those who had lost their jobs should take
up work normally done by foreign workers. They should not always
turn to the government to help. After decades of prosperity, and too
used to a paternalistic State, people found the advice that they should
buckle down and turn to making hotel beds and labouring on building
sites hard to swallow; so was the realization that real solutions to the
current problems might be out of the government's hands.

The economy has never totally recovered to the level of the prosperity of the previous two decades, and dangerously shows up the
fault-line of 'national enjoyment'. As for oral sex, it is not too much
to suggest that its excessive pleasure seemingly mirrors the unhealthy
propensity of 'national enjoyment' to get out of control, to create
over-expectation of what the State can and should do. Both oral sex
and public goods make the same point by the very enjoyment they
proffer. As one is in danger of spinning out of the orbit of moral
restraint, the other tends to become unreasonable and excessive by
falling outside how the State would wish it. The crime of fellatio as
'sex against the order of nature' may draw on the Victorian reasoning
of sex for biological conception. It also shows the depth of the State
anxiety and the need for regulation. In spite of the worries about 'human resources' and the infertility of graduate mothers, unnatural sex
is a subject of concern not because of these reasons. Fellatio is a perverse, excessive pleasure enjoyed in the realm of privacy and in lustful
self-regard. As such it is hard to tie this kind of pleasure to the 'culture
of gratitude'. The couple deep in the ecstasy of 'unnatural sex' do not
think of the State and what it has done for them. For the State with a
totalitarian ambition, 'national enjoyment' and perhaps all aspects of
social life can be invested with its own design and purposes. Yet the

very excessive pleasure of something like fellatio scandalously points to the opposite. Nevertheless people have to be made to remember, as they travel in the super-smooth Mass Rapid Transit (MRT) rail system and as they are being tempted by the goods of the trade union-run supermarkets (with discounts for members), where these have come from. Only when people remember the source of their enjoyment can they remember the story of the toil and triumph of nation-building by the selfless PAP leadership. The alternative is to give in to appetite: the self-consuming pleasure enjoyed to excess. And appetite signals all that the State abhors, because appetite is a form of pleasure that owes its origin to no one, that traces its contours on nothing except its own sensation and the bodies that experience it. In the State's view, fellatio is appetite par excellence. It has to be regulated because those who engage in it are obsessively and irresponsibly for themselves.

Conclusion – 'by way of the stomach'

In a short, enigmatic essay, 'Breakfast Room' (1979), German philosopher Walter Benjamin has us revisiting that strange dreamy state in the morning, when we are not yet fully awake and on an empty stomach. For most of us, the early hour before breakfast is a struggle:

> A popular tradition warns against recounting dreams on an empty stomach. In this state, though awake, one remains under the sway of the dream. For washing brings only the surface of the body and the visible motor function into the light, while the deeper strata, even during the morning ablution, the grey penumbra of dream persist, and, indeed, in the solitude of the first waking hour, consolidates itself.[26]

Against the soporific pull of sleep, the choice of being awake or staying in the dreamy state, and of taking breakfast too, takes on an existential import. A hearty morning meal foils the spell of dream. And those who seek to stay in the dream world surrender themselves to the 'protection of dreamy naïveté', and thus to 'bad faith' itself:

> He who shuns contact with the day, whether fear of his fellow men or for the sake of inner composure, is unwilling to eat and disdains his breakfast. He thus avoids a rupture between the nocturnal and the daytime world.... The narration of dreams brings calamity, because a person still in league with the dream world betrays it in his world and must incur its revenge. Expressed in modern

terms, he betrays himself. He has outgrown the protection of dreamy naïveté, and in laying clumsy hands on his dream visions he surrenders himself.[27]

From the far shore of clear-headedness brought about by bacon and eggs, washed down with a strong cup of *café au lait*, the enjoyment of the breakfast room looks like 'utopia':

> For only from the far bank, from broad daylight, may dream be recalled with impunity. This further side of dream is only attainable through a cleansing analogous to washing yet totally different. By way of the stomach. The fasting man tells his dreams as if he was talking in his sleep.[28]

But what is so virulent about dreams, and why does their rupture demand the strength and clarity that a good, hearty breakfast brings? One may read Benjamin here in terms of his injunction against the power of ideology as he outlines a mode of resistance. The play of metaphors, and the deploying of a mundane practice barely worthy of the name of philosophy seem to quietly lead us to something more insidious, something that stubbornly pits itself against the vigilance of our conscious life. 'By the way of the stomach' is a hint left casually so as to bring home the importance of *the senses* – and enjoyment in general – in the social and political realm. Benjamin is surely right; the dazed hours of the early morning would have captured what he must have witnessed with horror, the mass intoxication at the Nazi rally where under the light of a thousand torch lights bodies in unison dreamt the dream of national-socialism. In this and other instances, the collective pleasure of nationalism worked to the pitch by Hitler's spellbinding speeches was embellished with the designs of the Nazi State which delivered them directly to the bodies, so to speak.

Not only the totalitarian regimes, but all modern states in varying degrees make delivery of 'enjoyment' one of their priorities. From security of life and property, to social welfare provisions, patriotism and national identity, these are 'national enjoyment' at its most self-affirming and powerful. For the state, how handy if the external, social agendas are seamlessly at one with the inner life of the national subjects. Each experiences national goals as pleasure; each carries the 'national cause' within himself as a wakeful dream, Benjamin would say. The self-delighting subject in this mode is exactly how the state wants him to be. In the congruence of national agendas and personal feelings, enjoyment produces a kind of hegemony where the instinctive habits of daily life are imperceptibly merged with state power. The sensate body – and its inner desires – now finds a natural place in the purposes of the state.

For Singapore, however, the problem of 'national enjoyment' is perhaps more modestly about how to keep it within bounds, how to prevent it from becoming self-regarding appetite. If provision of the good life has been the mission of the State, even more so is its attempt to regulate 'popular enjoyment'. And the situation is never as simple, as the greater the social delivery the surer the support for the State, and firmer the ground for the building of State power. Especially in the condition of prosperity, people insist on greater enjoyments of various sorts beyond what the State would define as 'good for the nation'. Moving beyond the 'bread and butter' issues, the pursuit would take many Singaporeans to appreciate the arts, for example, for the enjoyment that this proffers. Sexual enjoyment, especially of the 'unnatural' kind, is clearly different from the comfort of well-provisioned public housing which is easy enough to be turned into electoral votes for the government. 'Sex against the order of nature' is out of touch with reproduction and electoral politics; it also tends to take the national subjects to the unruly, hedonistic pull of the senses. This kind of pleasure – appetite actually – is personal freedom, but, for the State socially, ideologically perilous. The State regulation of 'unnatural sex' may seem like a futile attempt to legislate conduct in the bedroom, but it sits well with its hegemonic ambition to bring all spheres of social life under the State's purview. Taking oral sex offenders to court does not so much protect society from moral calamity as assure the State that it can, and has the right to, intervene in the intense, lustful appetite people enjoy in private.

The current regime, under the younger leaders, is not unaware of some of these contradictions. The current policy shift is not about removing *in toto* the repressive measures no longer workable with the rising demands and new aspirations of society. In an exquisite 'halfway house', when some of them are being modified and rewritten the old spirit is being retained. There are some gilt-edged examples of this. There is currently much talk about the 'pink dollar' – money coming from serving gay tourists and local homosexual professionals. Real estate developers are putting up condominiums equipped with gyms and super-large showers to facilitate sex under the spray, so I am told by the real estate agent, in order to attract gay buyers. Money is to be made from the practitioners of the 'unnatural sex' that is still illegal. In a move to showcase its more liberal policies, the public service has announced that it will employ gay men. The rub is that they have to declare their sexual orientation, thus openly admitting to a lifestyle prosecutable by law. Gay men are members of society and deserve gainful employment like everyone, the Prime Minister has said,[29] and surely the wisest and the most humane thing to do is to repeal the Indian Penal Code that still occasionally hauls homosexual men to jail. But that the government will not do.

The State's willy-nilly suppression of 'unnatural sex' keeps company with other relaxations of policy. Yes, sale of chewing gum is no longer banned, but one can only buy it from the chemists, presumably for medical reasons. Singapore now has its own Hyde Park Corner where people can air their views and unhappiness with the State. But a police permit is needed, and what goes on there is not the free-wheeling affair of its London namesake on which it is modelled. It is, as they say, *plus ça change, plus c'est la même chose* (the more things change, the more they remain the same). What we detect in these more liberal measures is the need to hold on to the original ethos, the ideological 'deep structure' that cannot be compromised. In his eighties, Lee Kuan Yew is still active in government, holding the post of Minister Mentor. He may be too old to 'dash around, meet people, talk, dictate, get things done', as he said, but he will be keeping an eye on things and help in policy matters.[30] Regarded by many with reverence, Lee will keep the PAP government on track, to make sure that it sticks to the spirit of pragmatism, meritocracy and administrative efficiency. And 'national enjoyment', we may speculate, will be a central issue for the government under his son, the Prime Minister Lee Hsien Loong. The central political vision will still be this: as the State so comprehensively ensures the nation's happiness and prosperity, it must also, by right and by duty, keep consumption and expectation 'within reason'. As for the pleasure of the bedroom, it has to be regulated to thwart those self-possessing, recalcitrant national subjects – even if the State is not so boorish as to imagine that people would take such pleasure with the solemn practicality of 'national policy'.

6 'Talking cock'
Food and the art of lying

> ... there was an element of treason to common sense in the very objects of common sense.
>
> Saul Bellow, *Humbold's Gift*

Lying in their teeth

In Russia, as everywhere, people tell lies; then they tell a kind of half-lies – fantastical exaggerations of some elementary fact that challenge the credulous, and the Russians call these *vranyo*.[1] While lies (*lozh*) are downright untruths, told in order to deceive, imputing persons with the skill of *vranyo* is to say something about their lively imagination and ability to tell a fanciful yarn: 'You are having me on!' the listener will say, half in admiration. Ronald Hingley calls *vranyo* Russia's 'national brand of leg-pulling, ribbing or blarney'.[2] In Dostoyevsky's essay, 'Something About Lying', the Russian art is shown at its most subtle:

> Not long ago, I personally, while sitting in a railroad car, chanced to listen during two hours of the journey to a whole treatise on classical language. One man was speaking and all the others were listening.... [He] dropped his words weightily. It was obvious from his very first words that not only did he speak but probably, had thought about this theme, for the first time. So this was merely a brilliant improvisation.
>
> He emphatically rejected classical education, and its introduction into our schools he termed 'historical and fatal folly'; but this was the only sharp word which he had permitted himself. He had adopted too lofty a tone which restrained him from flying into passion, from contempt itself for the subject. The grounds on which he stood were most primitive, permissible, perhaps to a

thirteen-year-old schoolboy . . . to wit: 'Since all Latin works have been translated, Latin is not needed,' and so forth and so on, along these lines.[3]

The railway journey is a congenial place for such bantering, and the passengers impatient to arrive and with time to kill make a ready audience. *Vranyo* is thus also about social pleasure; it is 'almost invariably [told] for the sake of hospitality'.[4] In this form of lying, Dostoyevsky explains, '[one] wishes to create in the listener an aesthetic impression, to give him pleasure, and so one lies even, so to speak, sacrificing oneself to the listener'.[5]

Just as *vranyo* gives pleasure in the light entertainment and the company it makes among strangers, there is pleasure in its social function too. For that arch-patriot and novelist, lying is part of the 'fundamental traits' of the Russian people. Weary of truth, they need to see a favourable image of themselves, and '[crave] for instructors in political and social matters'.[6] These traits, to Dostoyevsky's mind, are cultural deficiencies of a nation lagging behind modern Europe whose civilization and technologies it had attempted to achieve for itself since the time of Peter the Great. In such a climate, a good teller of *vranyo* is always eagerly listened to. Through his elaborate tales and quick suggestions, he satisfies the national instinct for seeking 'instructions'. On their part, by abandoning themselves to yet another improbable rendering of 'the Russian problem' and its solution, the Russian listeners can bathe in the illusion that the problem is not as bad as normally thought, or its seriousness can still be undermined with jokes or cutting humour.

Later, in the twentieth century, *vranyo* was to have a special place in the Soviet Union. As it previously worked to compensate for the paucity of the Russian character, in contemporary times the Russian 'fantasy-mongering' eased the burden of living in a joyless totalitarian state. 'Indeed', Hingley writes, '*vranyo*'s continuing late twentieth century vogue may well derive its function of enlivening the drabness of modernity'; it remains relevant not least because the 'official doctrine of totalitarian Russia is somehow more exhilarating than life' and 'so extreme an example of creative fantasy'.[7] As in Tsarist Russia, *vranyo* is a particular response to the state. In easing the burden of living, the 'art of lying' in the modern Soviet Union took itself one step further. For when state planners announced a quantum leap in grain production from the previous year, or the building of so many hydroelectric dams, only the most ideologically zealous would see these as anything but official fairy-tales, told to glorify the achievement of socialism. They had all the shades of the age-old Russian art of *vranyo*. As people told their own, less flattering versions of the 'glory of socialism', they mimicked the state steeped in the subtlety of *vranyo*. In such context, *vranyo* was a rich fount of subversive humour when it poked fun at the official stories, when it mocked the grand gestures of the state.

We begin, improbably, with Dostoyevsky's meditation on the Russian art of lying for it offers a delicate reflection on a similar activity taking place every day in Singapore. Singapore's own 'art of lying' also takes the state as a theme from which people spin out splendid half-truths and rumours. *Vranyo*'s aesthetics of enjoyment, its public performance and social logic and, of course, its subtle inferences to the demands of the state: all these are found in the Singaporean practice of 'talking cock'. For Dostoyevsky the telling of *vranyo* best takes place in a train carriage where words and, one presumes, vodka flow as freely as the imagination; 'talking cock' is staged in the equally prosaic setting of a hawker centre or an outdoor restaurant in the cool of the tropical evening. In the conversation the triumphs and obligations of everyday living weave in and out of more solemn issues of state policies and news of government leaders, or whatever fancy chooses.

However, to speak of 'talking cock' as engagement with the State is to give the practice a formal quality it does not have. What lies at the heart of the Singaporean art of lying is precisely that imperceptible feature that so moved Dostoyevsky: the verbal improvisation that is the source of enjoyment, but also drawing on the motifs and tension of daily living. The nature of *vranyo*, as it is of 'talking cock', is both personal and public. The verbal inventions air repressed feelings, while they slyly 'answer back' to the state and its pressures. The visceral pleasure of 'talking cock' is indiscernibly tied in with the everyday with its hidden dimensions of power and significance.[8] When it takes the PAP State as a subject of improvisation, 'talking cock' becomes a part of what political scientists Gilbert Joseph and Daniel Nugent call the 'everyday forms of state formation'.[9] Linking the quotidian and the personal with the more broadly political, the concept takes us not to the state's structural features and awesome posturing but to its everyday effects and the retort from the people. In the 'everyday forms of state formation', we may say, people in Singapore contemplate the burden and enjoyment of life under the ruling PAP.

But 'talking cock' also gives pleasure of the more visceral kind – that of exuberant talk and masculine boasting, lubricated by delicious food and abundant alcohol. Here, conventional progressive thinking would have us see the enjoyment of 'talking cock' as the classic effect of mystification; it numbs awareness, replacing real political action with idle talk and leisurely enjoyment. Nascent political vision is drowned by the pull of the senses, so it can be argued. After all, the PAP State has always mixed harsh measures with efficient social delivery, so that what people enjoy in daily life – from the public housing to smooth traffic and the abundant goods at the shopping malls – comes to mirror, willy-nilly, the wise efficiency of the State. 'Talking cock', with its transparent, daily enjoyment, strikes precisely against this kind of 'leftist puritanism'. If pleasure and politics are indeed closely intertwined, then we have to take the grammar of enjoyment and the social formation of pleasure very seriously indeed.

Certainly the PAP State's effect on the everyday cannot be all repression; neither are people's responses all heroism and resistance. Like the passengers in the Russian train, those who gather at the hawker centre to take the evening air are doing it for the enjoyment of it and for imperceptible social reasons. 'Talking cock' through the self-affirming pleasure helps people to come to terms with an oppressive State that is also the source of economic security and social stability, and the symbol of nationalistic pride and identity. For its enjoyment, 'talking cock' is a platform on which people achieve a kind of self-knowledge on all these things. Instead of casting a veil over self-awareness and reflexive thoughts, the enjoyment helps people to manage the anxiety in their relationship with a state that punishes as frequently as it rewards, that exhorts both communal Asian Values and the virtue of heartless Darwinian competition.

'Eating air; talking cock'

For anyone who has lived in Singapore, and has friends who are not above letting their hair down occasionally, the ubiquitous event needs no introduction. The setting is the key. 'Talking cock' is a festive event of the evening, at the end of a working day. For blue-collar workers – and professionals who still find in such eating-places a certain exotic grit – the gathering point is most commonly the hawker centre. Singaporeans – and Malaysians across the Causeway – have a word for the wonderful respite from the heat of the day: *chi feng* in Chinese or *makan ingin* in Malay, literally meaning 'eating air'. Like the English phrase 'taking a constitutional', 'eating air' evokes a scene of people taking a stroll, stopping to greet a friend or two, and the walk may end up at the food stall or restaurant. So 'eating air', like the mid-day siesta in colonial Indochina introduced by the

French, is an adaptation to ease the hardship of tropical living. The enjoyment demands no strenuous effort on the body and mind. In this atmosphere, verbal exchanges take their own course and no one is to be reprimanded for telling a tall tale or two, as long as it is done interestingly and with flair.

So this is the setting that meets the eye: men sitting around a table where good food, congenial company and liberal glasses of beer oil loquacity and loosen the rust of inhibition. In a no less dramatic manner than what takes place in the Russian train carriage, tales of wayward children, the triumphs and failures of one's venture at the stock market, the pleasure of the massage parlour during a visit to Thailand and, not least, government policies and State leaders are spun out in their endless variations. The tellers aim to entertain and to show off their verbal virtuosity. And as with *vranyo* too, there is a detectable social import in the eager seizing of subjects and their elaborations. For by these elaborations, the wild tales of personal triumphs and hard knocks, of demands of the state and its deliveries, are transformed into ones that articulate the tellers' concerns over those things that deeply affect them.

Talking cock, as the *Coxford Singlish Dictionary* usefully defines it, is the act of 'speaking nonsense' or 'speaking rubbish'.[10] Probably derived from the English phrase 'cock and bull story', it refers to speaking half-truth, or at least an artful embellishment of some commonly recognized fact. On one level, Singaporeans use the term as an expression of astonishment when they are told of something hard to believe, such as, in colloquial 'Singlish', 'Ah Beng got promotion, ah? Don't talk cock!' Here the listener gently retorts, 'Surely this can't be true, and Ah Beng is such a lazy worker!' But the retort is also marked by something else. Once people see through a story as 'talking cock', then they do not take it as deception and its moral culpability becomes muted. The 'rule of the game' is not to counter-argue in order to extract the truth of the story – 'Has Ah Beng *really* got a promotion?' Rather it is to recognize the sad humour in the incongruity of it all: that an inept person, both at work and in life, could actually be promoted above the heads of more competent colleagues, flies in the face of the principle of rewarding work performance, and so on. The story is thus a form of teasing, of 'pulling one's leg' ('surely you can't be serious!') – even though it clearly implies a critique, as jokes variably do. As a gesture of fun and a source of critique, here 'talking cock' reveals its intricate character. For with these somewhat contradictory properties, the Singapore mode of 'lie-mongering' seems to undo itself.

It cannot help being funny, yet carries a quiet, sobering wisdom about the facts of practical life.

Meanwhile in Clementi New Town, the pleasant event that I participate in takes place around a table in an alleyway, a public thoroughfare actually between a food centre and the NTUC (National Trade Union Council) supermarket. There, the regulars – a taxi driver, a store man, a schoolteacher, a food-stall owner and I – meet to eat and drink, and to talk with abandonment. Any subject will do. This evening Ah Ng, the 35-year-old taxi driver, after meandering through other issues, turns to the high prices of cars, and the electronic road pricing (ERP) scheme that records and charges the number of car trips made to the busy Central Business District (CBD). Then, in a moment of classic 'talking cock', someone catches on and says: 'Ah, all the high prices, you know, they put us in debt so they can control us.' This is the conspiracy theory Singaporeans are so fond of: the State has brought the prices to exorbitant levels as a way of keeping people – especially middle-class professionals and business owners – in debt, so that they will be 'obedient to the government'. People working to pay the mortgage, so the common wisdom goes, do not leave their jobs to throw barricades in the street.

Meanwhile the sixth bottle of Tiger Beer arrives, as do other dishes of 'beer chasers'. Satay sticks, *puo pian* (sushi-like rolls of cooked radish and garnishes), *rojak* (spicy fruit salad with shrimp paste sauce) and, sometimes, barbecued fish in banana leaves add to the cluster on the table. Old Wong the stall-keeper sometimes brings over the chopped heads and entrails of spiced duck – leftovers from his Chiu-Chow porridge stall. They are only tit-bits; to make the meeting, it is best to have dinner before coming as no one seems to be interested in a proper meal of rice and cooked dishes. It is as though a relatively formal meal would be distracting; the demands on table manners and dexterity with chopsticks would take people from the pleasure of 'talking cock' for which they have come. So the small dishes arrive one after another, and men reach out for a morsel or two, and talk boisterously on. And here Ah Ng the taxi driver pauses, half for chewing, half for contemplating how to spice up the detail so casually dropped by the last person. Taking a sip of beer to lubricate his throat dried as much from eating as from talking, he adds to the chain of repartee: 'Yeah, those who earn good money have to be grateful to Lee Kuan Yew. But after all these years, now his family runs the country...'

'The Lee family runs the country'

In Singapore the idea of the country being 'run by the Lee family' often turns up in conversations – not least with the gregarious cab driver during the journey from Changi International Airport. It is at first hard to know what to make of it. Singapore has a parliamentary form of government with its basic rulings and constraints; the suggestion that Lee, even with his enormous political influence, can together with his family run the country single-handedly is simply not true. So what does the phrase mean? Does it say that the Lee family is able to bypass parliamentary scrutiny and the government machinery in making their decisions the basis of government policy? And if they indeed make such decisions, does state policy *ipso facto* reflect and advance their interests? These are however unfair questions, for we are dealing with a subject of rumour and hearsay, a subject of 'talking cock' more exactly. Truth or falsity is not really the point. Like lovers' praise of the real and imaginary virtues of their beloved, hyperbole is a rule of the game. And like lovers' praise too, 'white lies' or 'half-truths' tell much about the speakers, their desires and inner feelings. For this reason, for all its excessiveness the idea about the pervasive powers of the Lee family deserves consideration.

In the general election in November 2001, the ruling PAP returned to power having won all but two of 84 parliamentary seats and took 75 per cent of all votes cast. Compared with previous years, this new landslide electoral victory took place in quite different circumstances. The island was going through a hard time that perhaps recalled the bleak political and economic situation in the early years of national independence. Following the September 11 bombing of the World Trade Center in New York, Singapore had put itself firmly behind the United States and the war on terror. In January 2002, a plot to bomb the United States and Australian embassies was discovered by the Internal Security Department (ISD), and some 15 Islamic suspects were arrested. This happened at a time when there was 2.2 per cent shrinking in the economy and 4.7 per cent unemployment. The recession in Japan, decline in world – particularly United States – demand for its exports, and competition from China, India and Malaysia contributed to the economic downturn. For a nation which traditionally sees itself as an oasis of political stability and ethnic peace, and the shining star of the East Asian Economic Miracle in the two decades before the financial crisis of 1997, the effects of these events were traumatic.

The November election also made news with the impending change of leadership. The then Prime Minister Goh Chok Tong, who took over the leadership after Lee Kuan Yew stepped down in 1990 after 31 years in the post, announced that he planned to retire and would not be taking part in the next general election in 2007. The leadership would be taken over by the then Deputy Prime Minister Brigadier-General Lee Hsien Loong, Lee's eldest son, who later became Prime Minister in 2004. BG Lee, as he is called in Singapore, has been in the inner echelon of the PAP since he won a parliamentary seat in 1984. In 2001 he was Deputy Prime Minister, Finance Minister and the chairman of the Monetary Authority of Singapore (MAS), the de facto central bank, and deputy chairman of the Government Investment Corporation (GIC), the US$120 billion national 'fund manager' that invests Singapore's national reserve. Lee Hsien Loong's appointment was given much publicity by the government. Finance Minister is a post usually held by an older and more experienced man; the then 51-year-old BG Lee not only handled most of the financial matters of the State; he was also given the task of working towards a recovery plan for the besieged economy.

The idea of 'the Lee family running the country' also hints at the prominent positions of other family members. In the spinning of tales, men would also mention Lee Hsien Loong's wife, Ho Ching, who was until recently the chief executive of Singapore Technologies, Singapore's biggest unlisted government-linked conglomerate; she now heads the government-linked Temasek Holdings. Then there is Lee Kuan Yew's youngest son, Lee Hsien Yang, who leads SingTel, the nation's telecommunications giant and largest listed company with the government as the majority shareholder. And if further evidence is needed, the men hint darkly, there is Lee & Lee, the solicitors' firm of Lee Kuan Yew and his wife, which 'makes millions by doing conveyances for the sales of HDB [Housing Development Board] flats'.

As a subject of social bantering, the idea of the Lee family's iron hold on Singapore should be taken for what it is. But what of BG Lee's brilliant rising career, and why is it the fond subject of 'taking cock'?

When it was announced that he was to be groomed for the top government post, much was made of the fact he was a son of the elderly statesman and architect of modern Singapore. For the foreign press, this is evidence of Asian or Confucian patrimony where wealth and political power tend to be kept within the family, passing on from father to son. Such reporting is clearly a critical dig at Singapore's promotion of Asian/Confucian Values. Nonetheless like much of what happens in the modern city-state, such a crude, rigid understanding will not do. BG Lee may be the son of the senior statesman, but he has also precisely those personal qualities and educational qualifications which the State regards as critical 'leadership materials'. Like his father he went to Cambridge University. He graduated with first-class honours in mathematics and took a postgraduate diploma in computer science. After returning, he entered the Singapore Armed Forces in 1971 but at 32 left to enter politics with the rank of Brigadier-General. In 1979, he attended the Mid-Career Program at the Kennedy School of Government, Harvard University. Thus his career followed the path of that of many PAP leaders: educated in a prestigious university in England or/and the United States, a distinguished career in the public service or the private sector, having one's talent spotted by the PAP, and culminating in a fledging position in the cabinet before moving to top leadership.

With his education and brilliant public career, BG Lee would have found himself in the inner circle of the government even without having the ex-Prime Minister for a father. We have to approach the 'Lee patrimony' with a shrewder reading. Lee Kuan Yew is committed to meritocracy. Without the necessary personal and intellectual qualities, it is doubtful that BG Lee would have got PAP sponsorship even with his family connection. One wonders if the elderly Lee, given his temperament, would have brought any of his sons into positions of power if they had not measured up to the qualities he deemed crucial for political leadership. If Lee Hsien Loong's appointment smacks of 'the privilege of the right connections', as the men in the Clementi hawker stall imply, it is also a remark on the Lee family and the way their children have turned out. Lee and his wife have raised no wayward children who did not take to university and professional life: the children have not, as in more ordinary families, brought disappointment to their parents by insisting on striking out on their own by becoming poets or rock musicians. Considering Lee's belief in meritocracy and inherited talent, this would have been a tragedy of Greek proportions.

The State in a mask

Meanwhile the fortune of the Lee family lives in the popular imagination like some fact of nagging disquiet that cannot be shaken off. For the small company of men at the Clementi hawker centre, it is as though the fact weighs so heavily on their minds that it requires a particular way of telling, of bringing it out in the open from the dark recess of inner thoughts. A great deal of the pleasure of 'talking cock' lies in this, in letting out the anxious discernment in a setting of social enjoyment. Nevertheless 'talking cock' never loses its realism as it compulsively turns to the socially pressing. And these thoughts circulate around the table as freely as the food and glasses of Tiger Beer. From Lee Hsien Loong's rise to power to his brother's spectacular takeover of Australian telecommunications company Optus: these and other subjects are taken up with an eagerness induced as much by conviction as by alcohol. Sticking close to the facts one moment, then pushing them to the realm of baroque invention in another, 'talking cock' is pleasurable and socially gripping. When the verbal invention takes on the subject of 'the Lee family controls Singapore', the engagement with the State fuses truth with a large measure of the art of lying. Indeed the Singaporean art of lying seemingly embodies a theoretical wisdom akin perhaps to that expressed by Philip Abrams:

> [We] should recognize that cogency of the *idea* of the state as an ideological power and treat that as a compelling object of analysis. But the very reason that requires us to do that not to *believe* in the idea of the state, not to concede, even as an abstract formal-object, the existence of the state.[11]

To Abrams, the fearsome visage with which the state appears before us is an illusion; it is nonetheless endowed with unmistakable realism and believability; it is a mask that both conceals and confronts:

> [The] state is not the reality which stands behind the mask of political practice. It is itself the mask which prevents our seeing political practice as it is. It is, one could almost say, the mind of a mindless world, the purpose of a purposeless condition.... The state comes into being as a structuration within political practice; it starts its life as an implicit construct; it is then reified...and acquires an overt symbolic identity progressively divorced from practice as an illusionary account of practice.[12]

The reality of the state then is a result of its power and monopoly of legitimate violence, and no less of its 'triumph of concealment'.[13] When the appearance of the state is already grafted to the legerdemain that gives it life and authority, any engagement with the idea of it has to be similarly double-faced. For neither can we take the unrelenting realness of the state as mere mirage, nor should we be totally sold to its busy swaggering and threatening violence. Indeed, in Singapore the everyday view is likely to – literally – laugh at the presumptive, aggressive posturing of the PAP State. Yet people are forever aware that such posturing is backed up with a potent power of censure. With this sceptical evaluation, Singaporeans must take the State seriously – its status and the means of coercion at its disposal – and all the time *see through* its magic and illusion.

In a sense, we can see 'talking cock' as lampooning the grand gestures of the PAP State at every turn. For the men in Clementi, the PAP State's endless boasting of its achievement, of its wise and selfless leadership that has brought happiness to all, is 'talking cock' too. 'Everyone talks cock' is the cynical assessment at the table. For if Lee Kuan Yew is the Father of the nation who has worked tirelessly to turn Singapore into a First World nation, so the bantering goes on, he and some members of his family are not above accepting offers of prime real estate at discounted prices.[14] The PAP leaders may have devoted themselves to the betterment of the people, but they do this on enormous salaries with the Prime Minister receiving almost $2 million a year. The government's explanation is that such payment is necessary to attract persons of talent from the private sector; their financial sacrifices in entering public office have to be compensated. But the thinking in the street is likely to see all this in simpler terms. The official reasoning is just as easily evidence of the pervasiveness of 'talking cock' that affects even the State leaders: they too, like lesser men and women, are 'doing it for the money'.

So what goes on in the Clementi hawker stall looks increasingly like political engagement of sorts. With baroque irreverence, 'talking cock' airs nagging concerns and mocks the grandiose pretensions of the State. An event of the everyday, it is a kind of politics without form, a freewheeling account of state power without thought. What takes place has some of the qualities of early state formation in England described by Corrigan and Sayer.[15] In *The Great Arch* (1985) they describe the formation as a series of cultural revolutions, and their effects have come to be dispersed over, and then embedded in, daily routines and rituals and no less in the principle of government. Public rituals, patriotic songs and the national flag arouse collective passion, and foster Englishness and the idea of England as a national community. The state also shapes citizens into individuals of various sorts. Against the solidarity of citizenship, the state makes claims on the people in terms of 'distinctive categories (e.g., citizen, tax payer ... and so on)' and people too begin to think of themselves as different from others 'along the axes of class, occupation, gender, age, ethnicity, and locality'.[16] The individualizing effect, in this sense, helps people to work out their aspirations and enjoyment in relation to the state's various demands.

These dual effects of the state are also an important feature of Singapore life. Health care, public housing, education and economic prosperity build legitimacy for the PAP, and the State cannot do this without turning them into massive 'social enjoyment' for everyone. When the 'Singapore good life' is imperceptibly at one with the State, people experience political rule in a personal and visceral way – as enjoyment. For them the government is not some abstract thing made real periodically through the ballot box, but a pervasive presence to which is owed the daily gratification of life. Repressive, yet bringing social peace and material prosperity, the State and its formations are evident in the mundane, everyday routines: everything from the HDB housing and the crimeless streets to the cheap and delicious foods at the hawker centres is a visible sign of the State and its good works.

Back at the site of 'talking cock', we may perhaps ask: what of food? Satay sticks, baked fish in banana leaves and roast duck entrails, washed down with ice-cold beer, loosen the tongue and grease eloquence. Food and the rapid ebbing of beer bottles blunt the seriousness of 'talking cock' as social critique, yet warm the need to speak out. The verbal excesses, and the dazzling ability to turn the mundane into tall tales, PAP leaders into common seekers of profitable careers: all these are best accomplished when men are in an alcoholic stupor over an eating table. In such a moment, pleasure incites the compulsion to speak, and words irrepressibly speak themselves, as it were.

Pleasure: a political issue

If giving over to the compulsion to speak is a major enjoyment of 'talking cock', then we have to consider the emotional energy – the libidinal forces – that animates it. Pleasure shapes the tenor of excitability of the speakers. And in the midst of all this are issues – from the cost of living and price of cars to the fortune of the Lee family – that enliven the exchanges around the table in Clementi. Pleasure, excitability and engagement with the State are the cornerstone of the Singaporean art of lying. In 'talking cock' the visceral and the public, the realm of the senses and the garrulous affairs of power, merge and assume a fetishistic form.

We recall here Marx's famous dictum: 'Man is affirmed in the objective world not only in the act of thinking but with all his senses. The forming of the five senses is a labour of the entire history of the world down to the present.'[17] Of course, with things of human desires, 'social and historical influences' are only partially the determining factors. The problem is how to show up pleasure's insidious propensity to embody and hide in the senses, yet bringing into the equation Marx's great insight. Pleasure has a private and public domain and one writer who sharply underscores this is Fredric Jameson.

For Jameson the problem can be traced back to Adorno and the Frankfurt School with their classic ambivalence towards the 'culture industry' in capitalism. With a Marxist, liberatory agenda, Jameson could hardly make himself celebrate the enjoyment of the effusive culture of late capitalism without being reminded of the devastating effects of commodity on human consciousness. But Adorno's vision is clearly too bleak, and Jameson would see capitalism's pleasure as having a redemptive side as well. In 'Pleasure: A Political Issue', he writes:

> It is at any rate that the problematic of new revolutionary needs and demands and that of the commodification of desire and pleasure are dialectically at one with each other.... On a more populist view, indeed, the question might be raised as to whether all that mindless consumption of television images, that self-perpetuating ingestion of the advertising 'images' of things rather than the things themselves, is really all that pleasurable – whether the consumer's consciousness is really so false and so little reflexive as it dutifully treads the rotating mills of its civic responsibility to consume.[18]

When Jameson questions if the 'consumer's consciousness is really so false and so little reflexive', he gives the enjoyments of capitalism a degree of saving grace. We do not have to follow the orthodox 'leftist puritanism' that puts all pleasures of consumption under suspicion. However, Jameson is not willing to go along with celebration of consumption typical of postmodernism.[19] For Jameson neither 'leftist puritanism' nor postmodernism's collusion with capitalism will do; and his solution is to turn to the French critic Roland Barthes.

For Jameson, Barthes's *The Pleasure of the Text*[20] has a powerful lesson. It reminds the readers that 'literary languages' are often signs of power, and all literary practices 'symbolic endorsement of the class violence of this or that group against others'.[21] Barthes's solution is to advocate a kind of 'white or bleached writing' free from the utopian fantasy and longing that inspires class violence in the first place:

> In [the] attempt towards disengaging literary language, here is
> another solution: to create a colourless writing, free from all
> bondage to a pre-ordained state of language.... The new neutral
> writing takes its place in the midst of all those ejaculations and
> judgments, without becoming involved in any of them; it consists
> precisely of their absence.[22]

This neutral writing bleached of history does not reject literature's
social function, but it does rediscover a more sophisticated reason
for it. Writing of 'zero degree', as Barthes calls it, cannot be easily
harnessed to serve 'a triumphant ideology' or revolutionary cause.
Yet it allows the rewriting of history on a text free of the burden of
'original significance'.[23]

And this is how Jameson sees the 'political significance' of the unique
pleasure – *jouissance* – of 'text of zero degree':

> It is now through reception rather than production that History
> may be suspended, and the social function of that fragmentary,
> punctual *jouissance* which can break through any text will be
> more effectively to achieve that freedom from all ideologies and
> all engagement (of the Left as much as of the Right), which the
> zero degree of literary signs had once seemed to promise.[24]

And he ends by showing pleasure's inevitable link to the dilemma of
politics:

> [E]ven the flight from history and politics is a reaction to those real-
> ities and a way of registering their omni-presence, and the immense
> merit of Barthes's essay is to restore a certain politically symbolic
> value to the experience of *jouissance*, and to make it impossible
> to read the latter except as a response to a political and historical
> dilemma, whatever position one chooses (Puritanism/hedonism)
> to take about that response itself.[25]

For Jameson, the textual *jouissance* free of the nightmare of history
has a special power precisely because it is untainted by ideologies of
opposing persuasions. Text of zero degree resists becoming another
text of conventional political import, and in the process offers a carte
blanche on which readers can 'rewrite' their engagement with the
social and political surrounds.

Jameson endorses the Barthesian *jouissance* because it gels with his
vision of the socialist project and cultural-political criticism generally.

This vision has to do two things. It must describe and confront the bleakness of capitalism with unflinching realism. In doing so, it must sustain a degree of utopian hope and not give in to solipsistic despair. Progressive politics is after all about the imaginary future, and thus social investment and utopian hope. The trick is not to let the imaginary future provide solace for the oppressive conditions of capitalism.[26] In this sense, enjoyments of all sorts act like 'utopian gratification' by blunting the gruesomeness of the present. As pleasure incites the subject to endlessly return to the site of satisfaction, the present is made more liveable, more congenial, than a more realistic appraisal would allow it.

Power and everyday enjoyment

In Jameson's reworking of the pleasure of text of zero degree we begin to catch a glimpse of the burden of 'talking cock'. If this reworking shows up the subtle futility of the goings-on in the hawker centre, it is because they are underwritten by the double pleasures of the mouth. When people eat too much, when they luxuriantly give over to the 'talk' that estranges every subject from its normal meaning, they are caught in the pleasure of food as of their own eloquence. And this pleasure is always enjoyed *in excess*. As the evening drags on, the visceral experience binds them to their senses, rendering them oblivious to others and even the logical thread of their ruminations.

With rowdy self-absorption, there is no getting away from the fact that a large part of 'talking cock' is about expressions and discharging of pent-up energy. 'Talking cock' and 'eating air' give licence to let go, to bring into the open feelings and ideas about the State and its powerful effects that could not be expressed on normal, more sober occasions. Like the glow of post-coitus repose, the pleasure is more sensuous than cerebral. At the hawker centre, there is always a sense of men coming together for mutual confirmation, for airing their feeling of stressful ambivalence about living in a society that the PAP State has so deeply shaped. In Jameson's phrasing, 'talking cock' arduously seeks *solace* in the pleasure of the everyday that is, being what it is, free of 'significance' and barely worthy of note. And as solace, 'talking cock' delivers gratification in the face of people's powerlessness.

Still, 'solace' is not quite the right word. The pleasure of 'talking cock' belittles the past and distrusts the present in a social landscape in which the State looms oppressively large. Anything of a political issue and government policy the wisdom of the street quickly recognizes as already written by the PAP. Instead of 'nostalgia for the future' and socialist fantasy, 'talking cock' offers the quick satisfaction of derision. In the depth of alcoholic stupor, as dishes come and go, recollections of the salacious delight of the Bangkok massage parlour riotously meet reflections on the grand promises of the State and the wealth and power of the Lee family. The wild merging of subjects, the free marrying of the serious with the quotidian: these are the delights that have brought the men together. Derision cuts down to size the monumental pretensions of the PAP State and no less of the Lee family by placing them in the realm of gossip and lies and the pleasures of Thai prostitutes. If the result is to lighten the burden of living under the PAP, it achieves this by the satisfaction of unfastening the State from its powerful positions.

'Talking cock', like Barthes's text of zero degree, is potent of politics of a certain kind. Like text of zero degree, it allows people to 'reflect' on the effects of state power in daily life in a most extraordinary, least expected platform. It is a platform still free from official intrusion. For that reason some may even think of the 'hawker centre' as the best place to talk politics. Unlike the Speakers' Corner, you won't need a police permit and, should you get into trouble with the authorities, you always have the rowdiness of 'talking cock' as an excuse. Over time, the kind of 'engagement' with the PAP we witness at Clementi hawker centre has become a daily ritual of the evenings. 'Talking cock' offers a new freedom to think and perhaps to see life's possibilities under a PAP shorn of its repressive ways. In Clementi, there is talk of voting for the opposition in the next election and, perhaps out of desperation, quietly turning Chee Soon Chuan, the much-persecuted secretary-general of the tiny Singapore Democratic Party, into a political hero.

And as usual, the government is quick to catch on. During the tenth Parliament session in April 2002, some newly elected members raised concerns about what they called 'coffeeshop criticisms of officialdom' due to 'the government's failure to listen to the people'.[27] For them these 'coffeeshop criticisms' are admissions of the limited civil society and public debate about government policies. 'Talking cock' would be too indecorous a term to be used at the parliament. Nevertheless 'coffeeshop criticisms' lends 'talking cock' a certain formal dignity, and reminds the chamber that the State is not always the hegemonic centre of things, at least not at the hawker centre. Try as it may to define legitimate political debate as strictly belonging to electoral contests and the parliamentary floor, the State cannot but note what goes on in the coffeeshop and other places of everyday enjoyment. And what it notices there, as we do, is rowdy irreverence, sensuous enjoyment and, of course, subtle reflections on social and political issues.

'Talking cock' assaults government policies and the haughtiness of the Lee family; but what goes on is also gossip and rumour-making. And the gossip and rumour-making seemingly escape the hold of the State and its powerful censure. In the sites of the everyday, the enjoyment is animated by 'the hidden present, or the discoverable future', to use the phrases of French critic Maurice Blanchot.[28] And you need Blanchot's exuberant prose, rendered in English, to remind us of the elusiveness, the richness of it all.

> The everyday is a platitude (what lags and falls back, the residual life with which our trash cans and cemeteries are filled; scarp and refuse); but *this banality is also what is most important, if it brings us back to existence in its very spontaneity and it is lived* – in the moment when, lived, it escapes every speculative formulation, perhaps all coherence, all regularity.[29]

The everyday also escapes into the senses. Things of the everyday like the pleasure of food and 'talking cock' are not easily captured by abstract thinking let alone by state control; here lie their freedom and power. There is in these things 'some residue resisting analysis', and the sum of the experiences seems 'irreducible by human thought'.[30] Nonetheless, as we have seen, in the enjoyment of 'talking cock' currents of the psychological, visceral and political break their banks and flow into each other. The very enjoyment, with its exuberant verbosity and sensuous abandonment, re-links the private and the public, the personally pressing and the official measures. When 'talking cock' places social concerns in the unlikely platform of food and 'eating air', when men give in to the compulsion to speak in a way that defies the use of words like critique and resistance, it takes on a 'social realism' hard to ignore. For all its voluptuousness, the pleasure of 'talking cock' can still be socially engaging and subversive.

Conclusion: the double pleasures of the tongue

Eating and talking both employ the tongue: one for eating and the other for ingestion. For philosopher Jacques Derrida, the double use of the tongue is a sign that speech is much less logical and rational than Western philosophy would have it.[31] As speech is always touched by other functions of the tongue, and vice versa, 'talking cock' is to do this even more literally. Greased by food and alcohol, the kind of rumour-mongering that we have witnessed cuts short the *relatively* sensible and, we may say, logical verbal exchanges of normal circumstances. The luxuriant, excessive enjoyment turns the words around. Touched by anxieties, words spoken are not made insensible, so much as having their meanings changed and reshaped. In the daily gatherings in Clementi hawker centre, it is hard not to be impressed with the passion and certain truthfulness with which things are being said. 'The Lee family runs the country' is pure 'talking cock'. Yet people could not help talking about it in the way the tongue could not stop itself seeking the new filling after a visit to the dentist. 'Talking cock' is about the deeply psychological and the instinctive working of the repressed; it is also more than that. In Singapore the Lee family and the PAP are not something which people talk about in an easy and open manner like the weather or the prices of shares or delights of a New Zealand farm holiday. At a time of economic downturn, as stories of friends being laid off become common news, gossip about 'the Lee family running the country' reveals people's insecurity and the feeling that the good life of Singapore may well be a thing of the past. The gossip not so quietly casts blame on political leaders who have always boasted about their talent and achievements in bringing prosperity to the nation.

The Singapore art of lying, one might say, at once does away with history and yet deeply engages with it. Just as Barthes's text of zero degree is bleached of history, the pleasure of 'talking cock' detaches itself, as pleasure invariably does, from 'the political' by living in the immediacy of the senses. Yet ironically, in 'talking cock' the self-delighting subject also creates a fresh platform from which to scrutinize all that the State does and says. And perhaps 'coffeeshop gossip' is already an indictment of political life in Singapore. That State politics should be the obsessive subject of 'talking cock' surely makes for a lesson about the nature of political debate and the status of civil society on the island nation. As taxi drivers and school teachers do this in the hawker centre, middle-class professionals and academics have their 'talking cock' gathering around the swimming pool or in an Italian restaurant over bottles of Chianti, cheese and gelato. In these places, the urgency to speak digs deeply. As with Walter Benjamin's discoveries in the daily gestures of Naples where '[e]ars, nose, eyes, breasts and shoulders are signalling stations activated by the fingers',[32] a riotous freedom runs through the conversation. State policies are misread; the propriety accorded to PAP leaders is turned into snickering disdain. Matching the boisterous festivity of Neapolitan villagers, the charm of the Thai masseuse, in the quickening of the tongue, invades the respectful status of the Lee family and the fortune of SingTel. The cheap and delicious barbecued lobster in Bangkok market meets the high cost of living in Singapore and thus, by implication, the failure of the State in this regard. In these blasphemous meetings of the grotesque with the serious, one catches a glimpse of the political import of the Singaporean art of lying.

In the end, 'talking cock' is perhaps classically carnivalesque as famously described by Mikhail Bakhtin.[33] As Bakhtin might put it, the subversive potency of 'talking cock' lies precisely in transforming the tight instrumentalism of the PAP State into bacchanalian excesses. Prising open the State ideology, the relentless humour realigns the sterile calculations and constant anxiety, just as it registers an understanding of State power as violent and capricious. The carnivalesque gives 'talking cock' an undeniable, critical energy. It shows up the State discourse as sham; but it also reconstitutes new individuals not easily awed by the pomp and presumptions of the State. It subjects State power and life's issues to derision and, in the process, nurtures a sociality that has the feigned cadence of 'plebeian solidarity' and 'comradeship'.[34] In all this, lest we forget, pleasure is still something of the senses, something experienced for itself. The culinary enjoyment and the leisurely 'eating air' in the evenings pulls the carpet from under the feet of the normal fixtures of the State; people also do it for the enjoyment of it. Yet if the enjoyment mirrors the Barthesian *jouissance* of the bland yet politically potent 'text of zero degree', then 'talking cock' through its enjoyment and imperceptible ordinariness rewrites the Singapore Story and retells it as the people see it.

7 *I Not Stupid*
Localism, bad translation, catharsis

Finally, it is self-evident how greatly fidelity in reproducing the form impedes the rendering of the sense. Thus no case for literalness can be based on a desire to retain the meaning. Meaning is served far better – and literature and language far worse – by the unrestrained license of bad translators.

Walter Benjamin, *The Task of the Translator*

Good translation

My title refers to a recent offering of the infant Singapore film industry. First released in 2002, it has proved to be hugely popular, earning $3.2 million, more than three times the cost of production. Despite its popularity, it is not a good film. The characters lack subtlety, and the plot lines are somewhat predictable; it is in fact a relentlessly didactic film. But didacticism fits the theme like a glove, dealing as it does with the nagging concern of Singapore society: the State's excessive actions and their damaging effects on people's lives. The film is thus a political critique of a sort that directly confronts the State and its policies. Holding up the 'original story' of the PAP State, *I Not Stupid* tells it directly, without sophistry or Beckettian reticence. To the local audience, much of the viewing pleasure comes from revisiting the home ground, and hearing the tales of their lives under the PAP rule: pleasure, in short, that comes from the endearing accuracy of the narrative. A didactic note sits quietly behind all this. And the central message appeals to what people know so well: that the State is the 'cause' of the emotional sterility and oppressive anxiety one feels about life – even if it has also brought wealth and political stability. If *I Not Stupid* indeed depicts with startling accuracy the travail of life in Singapore, then we are made to rethink the nature of the 'critique' at the centre of the film. The fidelity of the film in treading the familiar ground, in retelling the 'original story' about Singapore in all its truthfulness, raises the difficult idea of translation. And translation is about 'the kinship of languages', as Walter Benjamin puts it, that conveys 'the form and meaning of the original as accurately as possible'.[1] But this 'kinship' of the original and translation is never straightforward; translation always tends to undo the 'form and meaning' of both. What then, we may ask, is the nature of a political critique that assumes that there is an 'original story' to tell, one that everyone knows and that can be translated truthfully?

I Not Stupid tells of the troubles of three young boys in search of good grades in Singapore's pressure-cooker education system. Kok Pin, Boon Hock and Terry are classmates in EM3, the lower end of the English–Mother Tongue classes for academically weak pupils. Kok Pin is a talented young artist, but he has to hide his interest from his parents, who prefer him to concentrate on maths and science subjects. Boon Hock comes from a poor family; obedient and reliable, he has to look after his younger brother and help out at his parents' food stall at the hawker centre after school. Terry, in contrast, is the son of a rich merchant, a manufacturer of the local delicacy *ba gua*, a savoury 'barbecue meat', who has brought him up with servants and all the material comforts. With a Dombey-like father and coddled by a domineering mother, Terry is a weak and selfish boy. Much of the drama among the three centres around Terry's failure to stand up for his friends when they get into trouble with the school.

Kok Pin's father works in an advertising agency that is Singapore in microcosm. The place is filled with the prevailing 'Singaporean self-loathing'. The parents of Kok Pin and Terry cross paths during a near car accident and later on when Terry's father, finding his *ba gua* business besieged by Taiwanese competitors, turns to the advertising agency for help. The campaign, at his insistence, is placed in the hands of the American creative director who has come to Singapore under the government 'foreign talent' scheme. The campaign is a disaster. Meanwhile, Kok Pin's mother is found to have bone cancer, and the school seeks volunteers to donate bone marrow for her surgery. Forced by his overbearing wife, Terry's father volunteers without knowing the patient is the wife of his nemesis.

Meanwhile the three boys also have their own adventure. An ex-employee of Terry's father, resentful at being sacked and for losing the money he borrowed to finance his trip from China, kidnaps the son but in a mix-up takes the other boys as well. The resourceful Boon Hock leads the break-out from the rural hide-out, and the kidnappers are captured by the police SWAT team. The courage and care of his friends change Terry from his pathetic tearful ways. After the hospital rejects his father's bone marrow because it is the wrong type, Terry for the first time defies his parents by insisting that Kok Pin's mother take his instead. Greatly moved by Terry's gesture, Kok Pin's father offers to take up the *ba gua* campaign, by modernizing the products to beat the Taiwanese competitors. Kok Pin's mother's recovery also makes her regret the harsh punishment she has meted out to force her son to excel in subjects for which he has no talent.

Produced on a budget of $900,000 by Raintree Pictures, the movie arm of the state-owned MediaCorps which runs most of Singapore's TV and radio stations, *I Not Stupid* has been since its first release in February 2002 a box office hit. It is the most successful Singapore film so far after the director Jack Neo's previous bitter comedy about Singaporean materialism, *Money Not Enough* (1997). I saw *I Not Stupid* with a group of friends one Sunday afternoon in Singapore, and felt out of place among an enthusiastic audience hugely enjoying the humour and the relentless dig at themselves and the government. There are a great many insider jokes in the film. The title in Singlish (colloquial Singaporean English); the crude, rich businessman too quickly turning conversation to swearing in Hokkien dialect; parental efforts to extract good grades from the children by caning them; a Western executive whose incompetence is only rivalled by the mixture of adoration and resentment of the local staff; the coddling mother – symbol of the patronizing, repressive State – who knows what is good for the children: these the audience instantly recognize as features of Singapore life. If these are stock figures and stereotype images, they are also endearingly familiar. They appear on the screen as if they have come straight from the homes, offices and schools. There is no doubt that this easy identification with the story is what makes the film so enjoyable for the audience.

Built into this identification, however, is another feature. As it portrays the pathetic endeavours of the adults, *I Not Stupid* chides them for their vanity and moral follies. The boys, as they struggle to cope with long hours of homework and suffer their parents, testify to the State's oppressive demands that have turned themselves and their parents into selfish, neurotic wrecks. There is no disguise of the didactic intent here. Reprimanding the adults is also to offer a critique of the State, with the children the innocent victims. Indeed when adult Singaporeans are shown at their most unsavoury and unflattering, there is no guessing where their values come from and what has made them so. Like a shadowy puppet master behind the curtain, the PAP State pulls the strings that animate the storyline. For the audience as much as for the film, the State and its repressive policies are too obvious to need mentioning, and they share this knowledge like some secret conspiracy. Identification and didacticism are pillars that structurally hold up the film; but, ironically, they also undercut the political critique it so loudly proclaims.

The film as allegory

To say that *I Not Stupid* tells 'the insider's story' for those in the know is to suggest that it is primarily an allegory. Like all allegories, say *The Pilgrim's Progress* or *Animal Farm*, it is a story of 'double meaning'; the story insinuates a deeper, more potent one of social and moral import. With allegory, one may say, subtlety is not the great concern. What drives the author is how to tell the story of human folly by deploying socially recognizable 'types' as main characters. In *I Not Stupid* hints are liberally given out so as to set up the characters and who they are meant to be. Terry, the overweight, pampered son of a wealthy businessman, not too subtly stands for the average Singaporean who has traded independence for security under a benevolent, authoritarian government. And the scene that illustrates this is familiar to many middle-class Singaporeans. Terry and the family are at the breakfast table; as he is about to dip the knife into the butter dish to butter the bread, his mother goes into a rage: 'Aiyah, how many times I tell you to leave the maid to do it.' Her point is that the maid is paid to do the work, and for Terry to help himself would merely upset the economic relationship. Having everything done for him, Terry symbolizes the Singaporean for whom personal initiative is an entangled choice involving, not least, 'political issues' about the State. And Terry's mother, dressed in matronly white, bedecked with jewellery, *is* the PAP State. Strong-minded and well-meaning, she knows what is good for Terry and his teenage sister – even if she has to whack them to drive home the tender message. Lest the audience misses the parallel, she is made to say more than once, 'You are so lucky to have a good mother who gives you everything.'

But it is the defiant teenage daughter who more fervently shows up the tension in the obligations and rewards of citizenship. Some in the audience will no doubt see her as speaking for those Singaporeans who, emboldened by education and economic mobility, are demanding greater personal freedom and political expression; she may even stand for the opposition parties. Having a mind of her own, she constantly comes up against her mother's interfering ways. In one scene, she fights against her mother's taste for 'rattan wall-hanging' and attempts to change the 'goofy' decoration of the bedroom. 'But this is my room; let me decide!' the daughter cries in frustration. She is also unhappy that she cannot make use of the *ang pow* – the Chinese New Year 'lucky money' for children – to spend on the teenager things she wants, since it is invariably put away for her. 'I have kept it and invest[ed] it for you', her mother says, echoing the practice of Chinese families across the island. The daughter replies with a sneer, 'Yeah, you keep my money and lose it in some of your stupid investments.'

For the audience, these and other scenes strike a familiar chord. Outsiders may complain of the tamely transparent characters; but to the locals what amount to stock figures and cliché situations are endearing and, it is irresistible to say, therapeutic. *I Not Stupid* retells the stories of their lives. For all its shoddy characterization and easy explanation, the film airs the grievances and quiet suffering people often keep in themselves. In this sense, the film is being slyly strategic when it chooses as its subject a widely shared topic of anxiety: the education system. The main scenes are set in school, and a large part of the subplots spins out of what happens to the three boys there. The incessant tests and examinations, the aggressive putting down in the schoolyard by fellow students with better grades, the sad comradeship of the three boys for being 'losers' in the school system, and the pressure at home to improve their marks – by punishment or employing a tutor: these scenes are scratching the surface of a more serious issue. For it is clear that the education system is not the real subject but that it is something else that casts a shadow on every aspects of social life – the PAP State. The school is merely another site where the effects of the State are present but which cannot be too openly talked about. Perhaps that is why *I Not Stupid* has something of the coffeeshop ranting about it. As with what takes place every day in the hawker centres, it cuts to the bone the gnawing concern that urgently seeks airing.

Thus it is the sad, bespectacled Terry who is made to carry much of the weight of the film. At home he is coddled by an over-protective mother. At school he is the schoolyard weakling who looks to his two friends for protection and comfort. When his friends appear before the principal for something they have not done, Terry, too easily breaking into tears, is unable to back them up as a witness. Fearful of defying the adults (read: government authorities), he is a burden to his friends when they escape from the jungle hide-out where the kidnappers have imprisoned them. In any case, it is his transformation from a tearful moral wreck to one who, against his parents' will, gives his bone marrow to save Kok Pin's mother's life that provides the melodrama. When he finally beats up the bully at his father's garden party, he has finally arrived at his own 'matured self'. The audience applaud when he presses a plate of fried noodles on the boy's face: a revenge of the nerd on all figures of authority. Terry is in short the victim of 'the government'. The message, if the audience have not got it already, is that the government makes decisions on most important matters, and its harsh and over-protective policies dull people's will and independence; the result is a society much like the infantilized Terry, suffering from the same debilities.

The 'foreign talent programme'

The 'pressure of the government' is not less severe outside the school. In the advertising agency, the trials and tribulations of the marketing of *ba gua* Chinese barbecue pork bring together the adults, and what happens reveals the shortcomings of a policy that affects many working men and women: the 'foreign talent programme'.

Like other rapidly developing economies, Singapore depends on foreign workers, currently about a quarter of the workforce. The majority serve the factories and construction sites, and foreign maids free women to work and support a lifestyle requiring a double income. For Singaporeans the presence of foreign workers is primarily a pragmatic matter. They are necessary for relieving Singaporeans of the one thousand and one jobs they cannot do, or that are too hard and poorly paid to be of interest to them. For the State, the 'foreign talent programme' is quite a different matter, however. It is designed to attract highly paid professionals and executives to Singapore. Most of these are from the West, usually transferred from senior positions in the headquarters of multinationals in Europe, the United States and Australia. In Singapore they enjoy levels of salaries and perks far above those of local executives often doing the same job.

I Not Stupid takes up the subject with a vengeance.

Terry's father, manufacturer of *ba gua*, commissions an advertising agency to re-launch the product and to halt the rapid decline in sales. When the campaign is presented to him by the local Chinese team, his outburst of profanity in crude Hokkien, all spitting and baring of teeth, makes all too obvious that he wants the best that his advertising budget can buy. He demands the services of the American executive. Earlier on the *ang moh* – to use the colloquial parlance, literally 'red hair' – is shown to have been promoted ahead of the more experienced local man; his Clouseau-like attempt to speak Chinese and to bite into a piece of *ba gua* is already a subject of ridicule and office gossip. For many people like Terry's father, no matter the *ang moh* is a stranger to Chinese culture and the local delicacy, being a Westerner he is simply the best person for the job. As it turns out, the marketing campaign is a failure; and it is Kok Pin's father who, with his cultural skill and local knowledge, takes over the project and averts it from disaster. Before that, the Singapore executives, in a moment of wounded pride, challenge the American to a competition to see whose presentation of a shampoo campaign the client will finally accept. When their plan is rejected, the Chinese executives lose their jobs as well: victims of the *ang moh* arrogance and the State's 'foreign talent programme'.

At the advertising agency, the grouches of the Chinese executives are at once about race and class. The audience is left in no doubt where the injustice lies. In the scene where the American first appears, he is made to fumble with the barbecue pork and, in a moment of cross-cultural empathy, exclaims its finger-licking goodness, 'It tastes good!' 'Well, if the job depends on it, I'd better learn to like it', he might have said to himself. Singaporean viewers would no doubt enjoy the scene showing the 'high and mighty' *ang moh* making a fool of himself. As one might say, since many have to 'suffer' smelly cheese and sandwiches during their European holidays, the shoe is on the other foot for this bit of culinary, postcolonial revenge. When he is promoted above the very capable local executive, and when he enjoys the confidence of Terry's father over others, the narrative reaches a pitch that cries out for bloodletting.

A stock character, the American is a figure against whom Singaporeans pit their unhappiness and resentment. At the same time, he serves to remind the audience that the true cause of the injustice lies elsewhere, in the state policy. *Ang mohs* are arrogant and over-paid – the men even enjoy the exclusive companionship of sarong party girls (SPGs), a contemporary variant of the 'sleeping dictionary' of the colonial days – and the 'foreign talent programme' has brought them by droves to the country. Again, *I Not Stupid* is timely, and strikes at the national psyche where it hurts. In the present economic circumstances, there is understandably a great deal of antipathy to-wards foreign professionals when locals are being laid off and new graduates cannot find work. Lee Kuan Yew, in an effort to settle the issue, explained why the government is not changing the programme. With the birth rate dropping to a low 40,000 per year, he reminded the tertiary students that finding 'world-class players among locals' is hard; and he set out the economic reasons why 'foreign talent' is crucial for Singapore:

If we do not attract, welcome and make foreign talent feel com-
fortable in Singapore, we will not be a global city and if we are
not a global city, it doesn't count for much. The days of being a
regional city, that's over.

There are four million people in Singapore; one million of which
are foreigners. You get rid of this one million and many will not
find jobs.

To those who said that the government should get rid of the programme to help local professionals, Lee went back to his fond subject
that people should be competitive and less dependent on the government:

In Hong Kong, where the Chinese went over when the Communists
came, you better find a job or, too bad, you get back to China.
Nobody owes you a living.

In Singapore, it's like the government owes you a living! That's
the way we developed but that's not the way we go forward.[2]

For Lee, the statement hopes to strikes home at a time when some of the
high-profile 'foreign talent' – including Danish coach Jan Poulsen of
the national soccer team – have recently left, and when the Singapore
economy is losing out to China, which is attracting some 70 per cent of
foreign investment to Asia. Not only do Singaporeans have to sharpen
their entrepreneurial skills by learning from expatriate professionals,
but they must change their mindset so that they are prepared to 'flip
burgers and make beds in hotels' when the worst downturn hits, as
people already do in Hong Kong.

However, the locals at the receiving end of the 'foreign talent' policy
are likely to see its effects in simpler and more personal terms. No
amount of economic logic and the rousing 'tightening the belt' will
make them forget the special treatment of Westerners they have to
put up with in the offices. The poor ethics is so glaringly clear that
I Not Stupid does not need even to whisper 'national interest'. As
Singaporeans like Kok Pin's father see better positions being passed
on to *ang mohs*, they experience the truth that life is ruled by unfair
competition. Lee's warning of Singapore's economy being hollowed
out by China, that people might have to return to those tedious jobs
long taken over by foreign labour, may arouse a greater pragmatic
sense in the people. Ironically it also confirms the need to be extra
self-seeking especially when it comes to dealing with 'foreign talent' in
their midst. To bear grudges towards them and sneer at their clumsy

ang moh ways may be ungracious, but more is at stake than good manners. To be rude and selfish may be necessary to get the best advantage in life. All these are brought up by *I Not Stupid*'s depiction of what goes on in the advertising agency. Singaporeans have a word for the uncertain feeling that is a mixture of jealousy, competitiveness and self-interest; they call it *kiasu*.

Kiasu – fear of losing out

In Hokkien dialect, *kiasu* is an adjective meaning 'afraid of losing out'. It is used as a pejorative term, referring to small-mindedness and unruly behaviour in getting the better of others. When Singaporeans describe it as their 'defining national characteristic', there is a sense of perverse pride and a self-mocking and grudging admission of the ugly side of themselves. There are spectacular examples of *kiasuism* in everyday life. From parents arriving at school in their Mercedes to 'fight over' the distribution of free school textbooks; to the piling up of food on the plate in a buffet – judiciously starting with the most expensive oysters and prawns – regardless of one's appetite; to the near riot crowd gathering at a McDonald's outlet after it announces the giving away of free gifts: these behaviours signify the almost pathological need to get something for nothing. As a moral fault, *kiasu* behaviour is ungracious, bad-mannered and motivated by greed.

Nevertheless the term can suggest a positive side as well. If the piling up of food in a buffet offends the civilized eye, it is also about putting oneself deep in the mud of the competitive game. Seen in this light, the aggressive shovelling of elbows at the table to get at the seafood looks like healthy self-regard. The State, of course, wishes Singapore to be a cosmopolitan city and the people gracious and civilized, and runs numerous public campaigns to achieve these aims. But the spirit of competition and self-interest always seems to makes a higher claim on people's behaviour.

Kiasuism is thus the subject of both social embarrassment and pride. For the need to always try to get the better of others also suggests, in another realm, one's consummate skill in the marketplace by beating the competition. With this ambivalence, one is not to be blamed but life's basic insecurity. For Singaporeans, the feeling is traced to the fast, cutthroat pace of work and the pressure-cooker school system and, finally, to the State itself.

This is really the driving theme of *I Not Stupid*. A mother constantly boasting of her child at the top of the class in order to slight the parents of less talented children may be a classic example of *kiasu* boorishness;

but she is not totally a character of scorn. She is merely being a proud parent, proud of her child and proud of having beaten the education system that casts lesser children by the wayside. On her own child, she may be quick to use the cane to make him or her study hard and get higher grades, but it is still for the good of his or her scholarly career. There are thus one thousand and one reasons for the *kiasu* 'national trait'. The mother boastful of her son and belittling of other children has simply been driven out of her wits by her worries; for Singaporeans she embodies their common fate and comes across as a figure of moral sympathy. The accusing finger points decidedly elsewhere.

In the end, the film is being disingenuous. As it teases the ugly Singaporeans, it quietly assures them that their ungracious behaviour is at least 'understandable'. It is not that the film explicitly approves of the *kiasu* behaviour it so sharply depicts, but just that people are living under a patronizing, authoritarian State which produces in people feelings of deep insecurity. Making furtive references to the State, *I Not Stupid* undercuts any serious critique of Singaporean boorishness and its causes. *Kiasuness* may be embarrassing and gross, but not a few are secretly proud of it because they are not, in the last analysis, wholly responsible for it. For Singaporean viewers, that the school system – and thus the State – is the cause of *kiasu* behaviour is merely common sense, a deeply felt truth of practical wisdom that contributes to the 'realism' of the film.

Realism and critique

An Australian academic friend saw the film in a cinema in Singapore and described his experience in an email:

> Looking fwd to hearing what you think of I not stupid. I saw it at Jurong East Shaw Bros in a cinema packed with heartlander families. V. parochial. At the start they actually DID clap then clap louder as instructed by the titles! But I've never seen an audience so engaged with a film here – no-one talking on their mobile phone during the screening. It was a bit like a giant group therapy session. Felt a little uncomfortable sitting there in my white skin (the only ang moh) when the lights came back on ... but no one wanted to lynch me.

Singaporean students watching the film on DVD over Foster's and chips in my sitting room find it no less engrossing. They too clap and cheer at the embarrassing, yet endearing, follies of Singapore life so truthfully depicted. If the film proves to be a 'giant group therapy session', it is because it offers itself to the Singaporean audience as a monumental 'in-joke'. No wonder that for my Australian friend the film is so flamboyantly 'literal' that he finds it hard to identify with it. Having seen the film, an outsider can be forgiven for having the impression that Singapore society is teeming with discontent, and the stress of daily living is bringing everyone to breaking point. This brings us to ask those questions that puzzle an outsider. In what way is *I Not Stupid* an expression of resentment against the State? And, indeed, what is the nature of its critical purchase?

Lest we have any doubt of the film's intentions, let us go back to it for a moment. After the title a line flashes across the screen in mock command, 'Now Clap!', and the audience actually do so in joyful unison. Again the joke is so chummy that people know instinctively what it is about: it teases them for their docile compliance with the PAP State. And the same 'command' is repeated at the end of the film, as one of the boys stares into the camera and shouts, 'How about some applause?' Here the audience laugh and clap knowingly too. There are other moments like these in the film. Terry's mother, accompanied by her two children, is shopping at the fruit stalls in the market. As they walk, the camera looks up at the mother dressed in regal white, with a bosom of some protuberance, and Terry's voice-over reminds us, 'Our relatives and friends say that she is like the government.'

Time Magazine describes the film as a kind of 'subversive underlying Id' to the 'censorious superego' that is the government. 'It's hard these days to find authority as all encompassing as in Singapore, where citizens learn from a tender age to watch what they say, do and even think.' 'With an unerring sense of place', the writer adds, the 'buzz that adroit criticism has produced shows that it may be time for the government to give the kids the keys to the car'.[3] For journalist Seah Chiang Nee writing in Malaysia's *Sunday Star*, the film marks a turning point of sorts; it is 'the first that criticizes – although indirectly – policies in Singapore since independence'.[4] In a similar vein, the publicist of the 2003 Seattle International Film Festival is keen to cast the film as a 'a sly and engaging critique of Singapore's obsession with losing its social, political and cultural identity to the twin "evils" of Western influence and modernity'.[5] In these views, *I Not Stupid* is notable for its patently critical position against the PAP State, disparaging it for the strain and anxiety it brings to everyday life and perhaps calling for greater freedom for the people to run their affairs. If the film does not intoxicate people with revolutionary fervour, it is still a 'critique' of a certain kind though it is hard to know exactly the nature of it.

Time Magazine is correct when it points to the 'sense of place' and 'timeliness' of the film. *I Not Stupid* was released in the midst of significant changes in Singapore. The PAP State was busily re-shaping the vision of the island's social and economic future. It is a 'telling point', one commentator suggests, 'that the S$900,000 film is produced by Raintree Pictures, the movie arm of government-owned MediaCorp.... It passes the censors without any cut.'[6] The Acting Minister of Information even praised the film for expressing 'people's sentiments about government policies'. 'We cannot say these polices are wrong because of the movie, but the movie reflected the feelings of the people and we should view such sentiments seriously', he said with some pathos.[7] Whether this is a sign of greater political openness or a post-Cold War Asian *glasnost* under the second-generation PAP leadership is a question involving a great deal of guesswork and po-litical wishful thinking. Many are more realistic. As one commentator says about the film's passing the government censor, 'The trends are changing but I'm not going to say it's becoming more liberal because of this one instance. The Government may become critical and shoot back over another issue.'[8]

We are on surer ground when we tie the film to the current concerns of the State. After three decades of high-technology manufacturing, precision engineering and science and biological research as the vehicle of economic growth, there is a general realization that these are not enough. A different model of development has to be found, and the State is beginning to examine the necessary changes for Singapore to compete in the new 'knowledge-based, entrepreneurial economy'. Perhaps towards this end the State is gradually opening up media debates about the importance of creative thinking, personal initiative and less rigid, spoon-fed learning for students. Arts – and popular culture like films – are regarded as crucial in nurturing 'innovation and creativity'. In a speech entitled 'Cultural capital and the creative industries', the Acting Minister of Information told the parliament:

> The creative industries will certainly be an important part of our future economy.... Beyond jobs, the creative industries also have a powerful, indirect impact on other industries – by adding style, aesthetics and freshness to everything we do, resulting in more exciting products and services. This will add vibrancy to our companies, and enhance the competitiveness of our industries.

And he went on with breathless enthusiasm:

> But being creative is not something we can order up on a menu. . . .
> [We] need to be stimulated by the buzz of a lively city to come up
> with original thoughts.
>   This is why cultural capital is so important. The richer our
> cultural capital, the more it nourishes us, and gives us that unex-
> pected insight or perspective to come up with breakthrough and
> winning ideas. . . . [T]he media industry is a huge global industry.
> In 2001, total worldwide spending in the media industry was $1
> trillion USD. We shall ride on this growth and work with studios
> to attract some parts of the value chain, such as digital animation,
> and pre and post-production work, to Singapore.[9]

So 'creative industries' are Singapore's new economic lifeline; and
being creative requires loosening up the political culture and pervasive
state control, and recognizing that unthinking popular obedience is
not a good thing. In this atmosphere, the State would almost certainly
approve some of the film's messages: that parents should not pressure
their children in rote learning, and that independent thinking should be
valued over blind submissiveness. In the new world of the 'knowledge-
based, entrepreneurial economy', the future 'digits of production', to
use one of Lee Kuan Yew's fond phrases, would be a generation of
creative, self-reliant and technologically savvy citizens.

I Not Stupid as translation

Looking at it this way, *I Not Stupid* becomes strangely complicit with
the State. When in the same speech the Acting Minister praised the
film for winning 'recognition at foreign film festivals' and saw it as the
beginning of Singapore's creative industries, one can almost hear local
progressives groan. Here again the arts, for all their diversity of forms
and exuberant personal expressions, are being quietly embraced by
the State because the new economy demands them. If a film like *I Not
Stupid* raises uncomfortable questions about the official measures, it
also points to the need for a spirit of independence so crucial to the
creative industries. For its harsh humour and relentless digs at the
State, the film betrays its critical sense when it insidiously finds a place
in the State's scheme of things.

    This is for me a compelling way to read the film. The logic of the
'new economy' weighs heavily on *I Not Stupid* and what it is trying
to say. There is, however, another problem, one about the narration

of the film, the way the 'Singapore Story' is told. With its accurate depiction of the tension and idiosyncrasies of life under the PAP State, we can think of the film as a form of translation. Translation seems the best word for describing what the film so richly offers: the direct and truthful rendering in *another form* of people's deeply felt experiences and their causes. Not only does it reflect how people feel about the education system, the 'foreign talent programme' and so on, but it also affirms what they know: the truth that the PAP State is responsible for the stress and unhappiness in daily life. There is pleasure in the latter too, by appealing to people's instinctive sense that the State is the 'original source of my problems'. We should ask here the kind of question Walter Benjamin has asked of translation: what happens in the process when meaning travels from one language to another? In our terms, what is gained, and what is lost, in the hands of the 'translator' who smoothly and truthfully renders the worries and contradictions of Singapore life in filmic form?

For Benjamin, the fallacy of 'good translation' is to make too much of the intimate 'kinship of languages'. 'Good translation' is not simply the transposing of meaning from one form to another:

> To grasp the genuine relationship between an original and a trans-lation requires an investigation analogous to the argumentation by which a critique of cognition would have to prove the impos-sibility of an image theory. There it is a matter of showing that in cognition there could be no objectivity, not even a claim to it, if it dealt with images of reality; here it can be demonstrated that no translation would be possible if in its ultimate essence it strove for likeness to the original. For in its after life – which could not be called that if it were not a transformation and a renewal of something living – the original undergoes a change. Even words with fixed meaning can undergo a maturing process.[10]

Translation then is always imperfect. The original meaning is 'infected' by the passage of time, just as it is altered by the interpretive intelligence of the translator and the people's reading habits, we may add. This has to be so because meaning in this journey can never retain its pristine state. The translator must confront the change in his enterprise. Ironically, meaning comes to be captured in its complex wholeness when the translator is not enslaved by the rule of 'faithful reproduction' or 'fidelity to the word'.[11] Benjamin's metaphor of the translator's task is not unlike that of the master Zen archer who aims with the mind's eye:

> The task of the translator consists of finding that intended effect upon the language into which he is translating which produces in it the echo of the original. . . . Unlike a work of literature, translation does not find itself in the center of the language forest but on the outside facing the wooded ridge; it calls into it without entering, aiming at that single spot where the echo is able to give, in its own language, the reverberation of the work in the alien one.[12]

'Good translation' is at best a flawed enterprise in this sense. 'Bad translation' breaks out of the literal, and the facile ambition of 'telling the story as it is'. Instead it attempts to capture the 'echo' and 'reverberation' of the original work by bringing into full play the translator's cosmopolitan sense and awareness of the changing nature of languages. And one cannot help but feel, to return to *I Not Stupid*, that the trouble with the film lies in its being too good a translator of life under the PAP. To say that is not to suggest that the realities of Singapore have been simply plucked out by the sharp, observant eyes of the director and transformed into images. What happens is much more deliberate. If the film is a faithful rendering of Singapore life, it is achieved by the cosy agreement between the storyteller and the audience as to the 'original story' and its significance. All the familiar scenes and the tension-releasing belly laughs point to this; they lock the audience and the film – and, imperceptibly, the Singapore State – in a warm embrace.

For all its critical ambitions, *I Not Stupid* tells once again the Singapore Story. What is this story except one about the PAP State's caring enterprises in giving the people economic security and perhaps a fulfilling life. Everyone knows the story without a second thought. Of course, the film is all blame and little praise in the treatment of the State; it nonetheless retells it with desperate accuracy. *I Not Stupid* transposes the omnipresence of the State with such force that it needs only to be in the background, providing the answers to all the questions, resolving all uncertainties. If the State has provided practically everything that matters, from prosperity to the smooth-running Mass Rapid Transit (MRT) trains, social peace to efficient public service, then it must be by the same token responsible for my disappointments in life, my mishandling of my children's poor performance in school, and my discontent at work due to the 'foreign talent programme'. Further still, even what largely amounts to ungraciousness in the ugly *kiasu* behaviour is, however tortuously, something of the State's doing. A performance artist in Singapore once complained to me, without a touch of irony, 'The government controls our practices – what we can and cannot perform; and it gives us very little funding.' Like the film, he so aims his critical ballast at the 'centre of the forest' that he no longer needs to ponder that interpretive wisdom might be found at the edge of the woods, as Benjamin might say.

Jokes and negotiation

Anyone who has flown to Singapore and taken the taxi ride from Changi International Airport to the city will remember the experience. Drowsy with fatigue and beclouded by the free alcohol onboard, one does not quite expect the lively commentary from the driver's seat. Singapore cab drivers, like their counterparts all over the world, are a gregarious lot much given to dishing out local information and gossip at the slightest encouragement. But whereas cab drivers in other cities talk of urban crime and lurid goings-on in the back seat, in Singapore they always seem to turn to politics and the PAP State. So, in the back, clutching your laptop and duty-free liquor, your are given the latest news about the dismal economy, the rising cost of living and the new appointment of Ho Ching, the wife of the Prime Minister, Lee Hsien Loong, to look after a government investment corporation. There is one world for the rich and powerful, and one for people like us, he says. Yes, Lee Kuan Yew has done so much for the country; now he has forgotten the poor people and is giving himself and PAP politicians millions of dollar a year in salaries. For a new visitor whose view of Singapore is about squeaky clean streets and the banning of *Playboy* magazine and restriction of chewing gum sales, the airy loquacity of cab drivers comes as a surprise. Perhaps the calm and political passivity is all a show, and underneath it are seething political discontent and potent radicalism.

Then the visitor checks into the hotel and, if he should venture out to meet Singaporeans, he finds that what we might temptingly call the 'Singapore cab-driver syndrome' is very common indeed. People are fond of complaining about the government. In the right company and given a sympathetic ear, they are driven to speak as if gnawed by an anxiety. But the grouches are more often about the practical things like the cost of living, the traffic charges and exorbitant import duties on cars and, of course, the stressful life in a fast-paced society; they are never about 'regime change'. Life without the PAP is simply not on the horizon of the collective imagination. With the weak political opposition deprived of resources and leadership, and the firm hand of PAP rule, change of government – and, God forbid, a Singapore ruled by the Workers' Party – is as wildly fantastical as the sun rising in the west. When Singaporeans say that they are being realistic when speaking of the PAP, they affirm the political status quo, the wisdom that the PAP is the only party that can effectively rule Singapore.

Meanwhile if all this sounds familiar, it is because a film like *I Not Stupid* is a symptom, a sign of the hard rock of the PAP State's hold on Singapore. Obsessively attributing to the State the 'origin' of my pain and happiness is the classic 'return of the repressed'. The State is as much a pressing reality as a phantom of excessive imagination that circulates in the way people think – and fantasize – about their lives. *I Not Stupid* with its meekly critical posture arguably smacks of resistance and subaltern heroism. And the humour too is seemingly subversive; the State in the figure of Terry's interfering mother is cut down to size by the ridiculous ways she is made to behave. Yet these are jokes of a particular kind. They are didactic and purposeful, aiming to strike at what people intimately know. As such, these are no innocent or pure jokes, but carry a hidden – and not so hidden – agenda. Freud would have called them 'tendentious jokes'; they are jokes of 'less value' because they have a secret intent, and have a specific audience in mind. An unsympathetic listener is in 'danger of being confused by their purposes or having [his] judgement misled by their good sense', Freud writes.[13]

I Not Stupid is burdened with a similar fate. As a tendentious joke, it cannot but be a monumental exercise in parochialism. 'Only jokes that have a purpose run the risk of meeting with people who do not want to listen to them', Freud reminds us.[14] Thus my Australian friend who saw the film may be forgiven for finding the viewing a somewhat 'mixed experience'. What he finds alienating, however, the local audience find engrossing. In the single-minded engagement with life under the PAP, the film strikes home two contrasting motifs.

I Not Stupid is primarily a joke for those in the know: this is the winning formula. Singaporeans clearly enjoy and identify with the film in a way an outsider cannot. When local viewers see the characters, they do not laugh *at* them because their excesses and boorish conduct are too much like their own. The characters are more humorous than ridiculous, and the viewers do not treat them with contempt. More cutting humour – or perhaps Monty-Pythonesque wit – would be needed to truly take the PAP State to task, to show some of the hollowness and insidious ideological purposes of its policies, in which case the film would probably find itself in the hands of the censorship board. As they are, the jokes are endearingly close to the hearts of the local audience and the designs of the State.

I Not Stupid as 'positive criticism'

In the end, the familiar story and easy characters render the film both pleasurable and cathartic. Blaming the PAP also gives intellectual clarity to people's unhappiness and its causes. With such intellectual clarity, as it pinpoints where the problem really lies, there is no need to ask deeper questions, to seek more arcane explanations of the dilemma of Singapore life. Like the cab driver who moans about the PAP, the film is in concord with the audience in giving over to the anxiety to speak, to unload the mental burden – so that they can return to the normal, 'repressive' course of life.

If the film is a criticism, it is what local people – and the State – would recognize as 'constructive criticism'. This is what makes the conjoining of desires of the State and the viewing audience. When the director Neo explains that his script 'was a constructive way of criticising' and that he 'wasn't doing it for political reasons',[15] he evokes a major trope of local political life. As the familiar refrain goes, 'the government welcomes public criticism, but it must be positive and not raise "sensitive issues"'. In making the ruling, the State is fond of raising the spectre of 'social harm' – like the race riots of the past – when freedom of expression is taken to the irresponsible extreme. But the rub is that the State is always the final arbitrator of what is 'constructive' or 'responsible'. This is in one sense the common ground between *I Not Stupid* and the State agenda. At the same time, the film also reflects the tenor of the popular responses to PAP rule. 'We recognise all that you are trying to do and we want to keep it that way; it is just that we want a bit more flex, a slowing pace', the audience may well be saying. There is in short an agreement, however grudging, that the State's overbearing measures are indeed 'for our own good'. 'Constructive criticism' does not question the fundamentals.

In this sense, *I Not Stupid* is typical of what passes for political criticism in the universities and the nascent civil society. 'Constructive criticism' is built on an acceptable platform from which hesitant critical voices can be raised, while nervously looking over its shoulder and guarding itself against breaching the limits. What happens is like the classic 'negotiation'. Imagine, if you will, two parties sitting across a table, brought together by an interest in ironing out the difficulties of some common issue, yet divided by their separate needs and interests. Negotiation, as a rule, unites and separates the parties concerned. It is not too much to suggest that the narration of *I Not Stupid* is like what takes place in the general elections that return the PAP term after term. Some people in protest against the State either choose not to vote or cast their votes for the opposition. But the electoral victory of the PAP confirms that popular discontent is rarely to question that 'the PAP is good for Singapore'. In the elections, as in *I Not Stupid*, there is something of 'dealing with the devil' about it. If people want the PAP to continue to rule, some social and moral costs have to be paid and, above all, excesses of the State have to be tolerated. As the film translates sharply the local feelings into another form, it also brings down the curtain to shut out the light of an alternative vision of how things in Singapore could be. A true critique would have to ask precisely the unanswerable: what would life be without/after the current political status quo? The localism of *I Not Stupid*, in giving pleasure to the viewers, blinds them with their own psychological obsession, moral earnestness and skewed view of Singapore's political future.

8 The nation after history

> The disappearance of Man at the end of History is... not a cosmic catastrophe: the natural World remains what it has been from all eternity.... Practically, [the end of History] means: the disappearance of wars and bloody revolutions. And also the disappearance of Philosophy; for since Man himself no longer changes essentially, there is no long any reason to change the (true) principles which are at the basis of his understanding of the World and of himself.
>
> Alexandre Kojève, *Introduction to the Reading of Hegel*

Sunday, 23 September 2001

The news comes through the mobile: there is going to be a memorial service tonight for the victims of the September 11 terrorist attacks on the World Trade Center. The Prime Minister and the then Senior Minister Lee Kuan Yew will be there, my friend tells me. So will thousands of Singaporeans. Breathless with excitement, he adds somewhat conspiratorially, 'The government is holding the memorial because the SDP [the opposition Social Democratic Party] is also having a rally tonight.' I tease him about his 'talking cock', as if the memorial service is meant to draw the crowd away from the SDP. In any case, government-organized events are as a rule massive and spectacular affairs, and no puny gathering of the opposition party can offer competition. We agree to meet. An initiative of the American Association of Singapore, the candlelight service is to be held in the National Stadium in Kallang at 6 p.m. We arrive an hour earlier, and a long queue of people is already there. Men, women and children of all ethnic groups – Chinese, Malay, Indian, Westerners – are waiting patiently to get in, many carrying American flags, while some are wearing T-shirts and ties with the Stars and Stripes printed on them. Outside the stadium is a small shrine. It is lined with candles, and covered with flowers, balloons, teddy bears, children's drawing of the American flag and letters of condolence to the victims' families. At 6, the crowd – there must be nearly 20,000 people – begin to move in, orderly and with hushed solemnity, and fill the stadium. Over the next three hours, we are taken through the paces, starting with the speeches by the then Prime Minister Goh Chok Tong and the US Ambassador, to the joint prayer for peace by Muslim, Christian, Hindu, Buddhist and Taoist religious leaders. Muslin clergy have a strong presence in the service. Mufti Syed Isa Semai, one of the most prominent Muslim leaders in Singapore, pleads to God to grant the 'planet eternal love, peace and tranquillity' and to 'make this world a safer one to live in'. The president of the Islamic group Jamiyah Singapore, Abu Bakar Naidin, also urges people to be 'united' and 'stand as one, showing sympathy for the victims' families'.

The Prime Minister speaks for less than an hour; he confirms Singapore's stand with the United States over the war against terrorism. Singapore shares the suffering of the American people, he says; the terrorist attacks in New York and Washington are attacks on 'humanity and the civilized world'. He praises the courage and determination of the American people, in showing 'great strength' and for 'being united and rallied behind their president'. Against the collective virtue and moral courage of the United States stands the absolute evil of the terrorists. They are determined and not afraid to die, and have no qualm in killing civilians and innocent people. For this reason, Singapore fully supports the use of military force by the United States 'until the driving figures are rooted out and their networks are disrupted'. Observing immaculate protocol, the Prime Minister has turned his attention to the suffering nation. But there is no doubt that terrorism and its devastating effects are also of grave concern to Singapore, if only because it is – as the refrain goes – an island of prosperity and peace in a sea of regional instability. Like other regional and world crises, this one too carries important lessons for the island republic. He ends by reminding the audience that, with what happened to the World Trade Center in New York a few days ago, it is necessary 'to reflect on the larger implications' and how Singapore itself 'should respond to terrorism'. And he concludes:

> The perpetrators of these terrible crimes must be brought to justice and others deterred from contemplating such horrific acts. We [in Singapore] will have regional and domestic sensitivities to manage. But we must accept risks for the sake of a better world.[1]

As we sit in the stadium with the evening light fast fading, the reminder of coming difficulties and 'regional and domestic sensitivities' puts a dampener on the promise of firm resolve and quick state action by the Prime Minister. With refrains of 'Amazing Grace' in our heads, we make our way home and conversation turns to the event we have just taken part in.

In many ways the memorial service is as people have come to expect of an event of this sort. Like the National Day Parade held a month earlier in August, this too is a spectacle of government presence and a rousing display of national solidarity. It is, above all, confidence bolstering with its assuring speeches and promises of firm policies. The theme of the National Day Parade was the 'battle' against economic recession and unemployment; of the memorial service it is war against terrorism. The PAP government delivers; this is what the people have come to expect. Since independence, the PAP has taken the nation through crises from the Communist-led labour unions and racial conflict to the threat of economic bankruptcy following the British withdrawal and later the separation from Malaysia. There is no reason to think that the government will be less determined and successful this time. My companions who have joined me in the memorial service are no zealous PAP supporters, but affluent blue-collar workers who can be disparaging of the State when it comes to issues that move them. The September 11 attack may belong to the volatile politics of the Middle East and religious passion from the mountains of Afghanistan, but for my companions it is another episode of the 'national crises', another play of the monumental problems and quick solutions on which the PAP government has staked its political reputation. History, both in living memory and as retold by the PAP, is the template for understanding the present. With history as the guide, my companions' near-absolute confidence in the PAP government's ability to 'fix' the current problems too is sound and sensible.

War on terror

Over the following months, the government began to actively pursue what it had promised to do. On 11 October, the authorities announced that they had discovered 'a plot by a Middle Eastern terrorist group to recruit Singaporeans to its ranks' in order to beef up its activities in the region. Two months later, 31 members of the Indonesian-based Jemaah Islamiah and other militants were arrested and detained under the Internal Security Act for planning a series of truck-bomb attacks on diplomatic, commercial and military targets in the city. Security operations gathered pace through 2002. On 8 September, the government arrested and detained 12 Singaporeans and one Malaysian for allegedly plotting to bomb US and other foreign targets in Singapore. After the lethal car bomb explosion outside the JW Marriott Hotel in August 2003 which killed 14 people including four Singaporeans, the government released a 50-page White Paper about the security situation in the region. Jemaah Islamiah (JI) is to Singapore as al-Qaeda is to the US and the world, the White Paper surmises: it poses the most imminent threat to the security of Singapore and the region. JI and other extremist groups including the Moro Islamic Liberation Front fighting for a separate homeland in the Philippines form 'a loose but trusted brotherhood of militants' over much of Southeast Asia. There can be no doubt of its dangers and, even 'if the US succeeds in dismantling al-Qaeda, radical Muslim groups in the region will continue to pursue al-Qaeda's agenda of global jihad', the report concludes.[2]

Singapore's quick response to September 11, by rallying behind the United States' war on terror and the arrest of Islamic militants, made news and assured the nation and its allies that it was leaving no stone unturned in seeking out the enemies of freedom and social peace. Words were backed up by actions. In October 2003, Singapore announced that it would be sending 192 military personnel, including a landing ship tank (LST) and a C-130 transport aircraft, on a two-month mission to join the US-led coalition forces in Iraq. Looked at from Washington and London, Singapore's small effort was impressive because it was not bogged down by 'domestic sensitivities' as in neighbouring Malaysia and Indonesia with their large Muslim populations. When the government banned the wearing of *tudang* headscarves by Muslim girls in the schools, it went ahead smoothly, and without the fuss and agitation there were in France when it introduced a similar ruling. Still, the terrorist threat could not have happened at a worse time. The three years from 2001 to 2003 were difficult times for Singapore. The economy was in recession with a downturn of global and especially US demand for electronic goods and other exports; and September 11 dashed hopes of a quick recovery. In 2002 there was SARS (severe acute respiratory syndrome). With the economy and employment still struggling to regain their former levels, SARS was estimated to have cost Singapore $1 billion mainly in tourism and pushed unemployment to a record high of 5.5 per cent.[3]

In the face of all this, the State promised to introduce new initiatives, and cajoled the people to be flexible and innovative to meet the changing economy. Yet there is a feeling that the times have irreversibly changed. Something has happened to the national psyche. When work takes me back to Singapore, I find it hard to figure out the new, deflated mood of life there. There is still the same professional swaggering among my academic friends ('When are you going to give up your pitiful wages in Sydney and come and join us?') and consoling nationalism of others ('We are so lucky that we don't have the same troubles with Muslims like in Indonesia'). But the realization that things can never be as good as before is never far off. Lim Kiat, a 45-year-old store man recently laid off after 15 years with the European furniture giant Ikea, now drives a taxi, and no amount of retraining is going to help to get his old job back, he tells me. He thinks of himself as lucky that he can find work while many of his friends are unemployed; driving a taxi will do for the rest of his working life. One of my Singaporean students, after graduating with a degree in economics, did a three-week coffee-making course in an Italian bistro in Sydney in preparation for the tough time ahead when he goes back. Most touchingly, my friend Daisy whose son is soon to graduate in accountancy from the University of Technology Sydney (UTS) feels that it is best that he stays on in Australia; Singapore is too small, and it no longer offers the same opportunity for young people as it used to.

Lee turns 80

I am moved to console my friends. Singapore's current woes are in many ways familiar to us living in Australia. Not only has it even more firmly put itself behind George W. Bush's enterprises in Afghanistan and Iraq, but Australia has its share of Islamic terrorist cells. After the bombing in Spain, many of us thought that Australia would be the next target of al-Qaeda vengeance. As for the economy, what is happening in Singapore is simply in the nature of a matured economy, I tell them. A certain level of unemployment and a modest growth rate are the costs of having arrived as a First World nation. The difference is that, while countries like Australia have accepted the situation, the Singapore State persists in telling the people that the current economic woes can be corrected – if only people will roll up their sleeves and be dynamic and innovative. When in November 2003 Singapore experienced 5.5 per cent unemployment (the OECD rate was 7 per cent for the same period[4]), it deeply traumatized the nation. After decades of growth, and the constant exhorting by the State that anything is achievable if people work harder and smarter, the economic woe is like a fall from grace. An acceptable level of unemployment, and slowing growth as a structural feature of a mature economy, is not the language for Singapore.

The economy is one thing; September 11 also brings home the realization that violence and terror are now of a global scale from which few – especially the open allies of the United States – can easily escape. In the past, security threats were the domestic and regional kind: the Communist insurgency during the Malayan Emergency (1948–60) and Indonesian commando raids during *Konfrontasi* (1963–66). The candlelight service at the memorial may have been a public display of support and sympathy for the United States; it was also a collective recognition of the danger of terrorism in Singapore, a public ritual that firmed up the feeling of the end of an era.

Still it is not all pessimism and doom. In 2003 the economy moved into a more optimistic gear, registering a modest average growth of 3.5 per cent, though unemployment remained unacceptably high at 5.5 per cent. There was also talk by the government of moving beyond high-technology manufacturing – computer parts and petrochemicals – which was facing increasing competition from China and India, and of staking a claim on the 'creative industries' of computer-aided design and animation. Singapore wanted some of the billions of dollars from the making of music videos and films like *Antz, Toy Story, Moulin Rouge* and *The Matrix*.

The year 2003 offered another bright light on the horizon of gloom. On 16 September, the then Senior Minister Lee Kuan Yew turned 80, a remarkable man at a remarkable, ripe age. The national celebration added some lustre to the dimmed hopes, and lifted the spirit of the nation. In the Shangri-La Hotel, some 1,000 guests, grassroots leaders, foreign businessmen and government officials gathered to honour the man. Hair thinner and greyer, and evidently having lost weight, Lee went up to the podium to thunderous applause. In his speech, he remembered fondly how life had turned out for him and the nation he had led:

> I cannot say I planned my life. That is why I feel life is a great adventure, exciting, unpredictable and at times exhilarating.... To make life worthwhile, never lose that joie de vivre. At the end of the day what I cherish most are human relationships.[5]

For all that has been said and written about him, Lee casts a giant shadow in the region remembered for its Marcoses and Suhartos and their cronies and corrupting power. His insistence on personal integrity and morality standards for his ministers produced no secret Swiss bank accounts, public statues of him in heroic poses, or grand edifices as monuments of Singapore's collective glory. And none would imagine that at his age Lee would be less robust in spirit. Nonetheless after 40 years of political power, it was a mellower Lee that people found that night. One former minister said of him, 'I find him more sociable and friendly as he moves around greeting old friends and colleagues, finding out how they and their families are getting on. The caring side of his nature is now more evident.'[6] In any case, Lee has left an extraordinary legacy. With his eldest son now taking the reins as Prime Minister, and with the team of hand-picked PAP leaders and civil servants, Lee can rest assured that Singapore will continue to be served by an efficient and incorruptible government. September 11, SARS and the economic recession will recede to be other crises that came and went.

The birthday dinner was a moving event. One is touched by the passage of time, the remembrance of the past that had been Lee's youth and the nation's beginning, and by the future of a Singapore without his firm guidance. The *Straits Times* reported:

> As the celebrations came to an end and the memories of Lee Kuan Yew burned as brightly as the candles on his birthday cake, Lee could not help but speak of the future. 'What will Singapore be like in 10–20 years from now?' he mused. 'We do not know how the cards will fall. There is always that element of luck.'
>
> He ended the celebrations by toasting the health of Singapore and all Singaporeans, choking back the tears that welled up just before he recited the national pledge. His voice broke toward the end as it became charged with emotion. It was a symbolic end to his birthday.[7]

'Singapore needs a break'

Lee's eightieth birthday also brings up a question in many people's minds, though the dinner was not the proper place for it. Having done so much for Singapore, and nurtured a competent team of leaders to carry on his legacy, is it not time for him to retire? Lee quickly put an end to such speculation. He announced that he would be staying on as Senior Minister in the Prime Minister's Department (his eventual title is Minister Mentor). To keep his seat in the parliament, he would continue to contest in the general election. 'I will retire from office when I am no longer able to contribute to the Government', he said; when that day will be 'depends on my DNA, my doctors, and the value of my data bank'.[8]

As usual Lee is one step ahead of the public thinking. In a special interview with the *Straits Times*, questions were put to him directly:

'Do you see yourself ever being able to withdraw completely from government and politics? When will you retire?' Lee's answer began with a firm rebuttal; he then gave his reasons:

> Your question is wrongly phrased. It is not government or politics I am involved in.
>
> I got into politics and government because I wanted to change society and bring about a better life for the people, and give their children a brighter future.
>
> I undertook this responsibility after I won the first elections in 1959, took them into Malaysia in 1963, and took them out of Malaysia in 1965. I still feel a responsibility for them. I can leave office, but emotionally, I will always be concerned about the future of the people of Singapore.
>
> I will retire from office when I am no longer able to contribute to the Government. But as long as I am fit and able, I will stand as an MP.[9]

The journalists persisted. Emboldened by the changes signalled by Lee's ripe age, their diffidence and awe for the man did not turn them away from the questions many people were asking. 'How do you ensure your presence and your strongly-held views on so many important issues do not cramp their [the young leaders'] style and allow them to experiment with their own ideas?' They continued, in a tone almost rhetorical: 'Why does it require the Senior Minister to suggest [new policies]? Why not the younger leaders? How come you are still the sole visionary?'[10]

These questions Lee answered with his usual directness and practical reasoning. Yet one detects in the interview a certain disquiet, a sense that his answers no longer satisfy, even for a newspaper renowned for its pro-government views. The pursuit of issues – of Lee's retirement, of his continuing influence in the government – suggests a certain daring and confidence in young Singaporeans. Can we read into the interview, if the journalists could come out and say it, the hint that Singaporeans would really like the Senior Minister to retire, and that his staying on is a disappointment? And what would they say about 'a Singapore without Lee'? How would they contemplate the future as the National Father gradually recedes into the background?

However, the *Straits Times* is not given to the confrontational style when interviewing government leaders. These questions are worthy of more trenchant reporting, but not for the *Straits Times*. For a more gutsy assessment of 'Singapore after Lee' we need to turn to the conversations and exchanges at dinner parties or at the hawker centre. And these begin with the perfunctory 'Lee Kuan Yew has done so much for Singapore'; then people go on to express – one can almost hear them sigh – that it is time for Lee to retire. Expressive of impatience, what they actually wish for is not immediately clear. It is certainly not about getting rid of a 'tyrant' and a yearning for political change or even a weakening of PAP rule. It is not even about a broadening social vision struggling to break through Lee's pragmatism and the State's push for economic priorities above all else. In most people's mind, Singapore's prosperity is too bound up with these things to have them radically changed.

Instead the general sentiment seems to be that 'Singapore needs a break (from Lee)'. For the academics and professionals I know, discussion invariably takes them to the changing economy and the flexible, more relaxed work practices they wish to have in Singapore. They have learned these in their study leave in the United States, or during their postings in London and New York. At the universities at home, they face a management style of concentrated power in a few individuals 'close to the government', so it is explained to me. These things are a part of the PAP way; they will not significantly change with Lee in the government where he can influence cabinet decisions 'with a phone call', as the journalists sent to interview him intimate. In a way, the wish to see Lee go is a call for a breather from the 'pressure-cooker conditions', to use the local phrase, in the schools and the workplace. For the professionals, experience overseas shows that open management and a relaxed approach to work and life do not harm efficiency and personal drive. One advertising executive, having spent six months in a New York head office, speaks with great fondness of the gathering near the drinking fountain in the morning, and the basketball court next to the staff canteen where staff can throw a few shots 'to help the thinking juice flow'. The longing for what goes on in places overseas, and the suspicion that work and life do not have to follow the Darwinian 'law of the jungle': these are the stuff that fuels the longing for changes at home.

The end of history

Yet it is never quite clear when talking to Singaporeans how the stressful pace of work and life is actually related to Lee's firm guiding hand, and how it will shift to a lower gear when he retires. To think that London and New York are more relaxed than Singapore can be a fantasy, I remind them. So there seems to be a great deal of grasping at straws in the thinking. In the end the wish for Lee's retirement may be driven by nothing more than the general feeling that the end of an era has indeed arrived. All that has happened in Lee's long and extraordinary life needs to be put to rest in order for the nation to get on to the next phase. It is like drawing a line at the end of a paragraph in a school assignment, or turning over a new page in a book after reading the previous one. The passing of the last half a century of national life shaped by Lee's indefatigable vision has to be given social and symbolic recognition. Perhaps that is why Singaporeans greeted the news of Lee's birthday in the way they did. It swelled from the deep-seated feeling about time's passing, about the coming of a new epoch registered by September 11, the economic recession and no less the 'intimation of mortality' of the National Father. They are, if you like, about the end of history.

The end of history – we need the theatrical phrase to give shape to what many Singaporeans are struggling with – does not have to be a grand statement about the death of communism and the triumph of liberal-capitalism. Neither does it only articulate the demise of the great ideological contest between the US and the Soviet Union after the Cold War.[11] People in daily life also have a sense of history evoked by the first landing on the moon, the arrival of electricity and telephone in the villages, the founding of a new regime after the revolutionary war and, of course, the death of a Stalin or a Tito or a Mao. Events like these are a signpost of change, a borderline on the 'time map' across which the old and the new traverse. For people of the new states, there is nothing that more sharply signals the *fin de siècle* than when tanks of the revolutionary army crash through the gate of the presidential palace and, after the euphoria, the sober task of nation-building. History may not be the word people use, but the feeling of hope and new possibilities is indelible. The hope and possibilities sometimes turn into ashes. As we encounter enough in the past, the ways of the coming regime can be the old horrors in a new garb, a barely disguised rehearsal of the old repressions: here ironically history is made new by repeating what went on before. This can also be the fate of history's ending: the nostalgic embracement of past legacies.

In Singapore there is no comparable violent revolution and grand bloodletting to reconstruct the past – even though the nation's brief history is retold as struggle, as filled with the traumatic events of its birth. Speaking of the present, the years after September 11 have proved to be equally unsettling. The National Father reaching 80 is by any standard a finale of a sort. When Singaporeans wish for Lee's retirement, they are not being 'ungrateful' nor do they envision the nation without the PAP. Rather it is to register the change of time, and with this a new reckoning that perhaps things can be done better and differently. In this sense, Lee's birthday does act like one of these critical events that evokes a new, albeit modest, vision of Singapore's future. Here then is an occasion to rethink a different balance of work and leisure, of political censure and personal and intellectual freedom. And for the professionals I interviewed, the new vision is precisely what they see in abundance in Sydney or New York: a more relaxed lifestyle without the spectre of social and economic collapse, contrary to the PAP State's vision. As with the journalists sent to interview him, there is no question in people's minds of Lee's continuing influence in the government, even without his holding a cabinet post. To wish to see him retire is to wish for his letting go of his 'guidance' when the younger leaders attempt to move to new policy directions more reflective of the global changes and more responsive to the new collective demands.

As Lee's birthday calls for national celebration, it also seemingly releases the suppressed boldness of visions, and the urgency to redefine the national ethos in people's own terms. The ageing of a National Father in any situation would indeed do this; his actual demise would do still something else. Continuing with the family metaphor, both eventualities signal the triumph of a child's Oedipal struggle with the Father, and a new dawn for the nation's maturity and identity.

The State at the end of history

Not only the people, but State leaders too feel the end of an epoch with Lee's birthday. The PAP is not, as I said, given to building statues and public monuments; nevertheless Lee's birthday called for a commemoration of some sort. At the birthday dinner, the then Prime Minister Goh Chok Tong announced the founding of a Lee Kuan Yew School of Public Policy. The purpose is 'to establish Singapore as a global point of reference for the study of policy and administration'. In collaboration with Harvard University's Kennedy School of Government, researchers will 'tap into Singapore's reservoir of experience and ideas, having transformed itself from Third World to the First', he said.[12] There is also to be a three-part documentary on the history of Singapore to be co-produced with a 'prestigious foreign partner'. 'The aim is to tell the Singapore Story in a way that will inspire Singaporeans to continue with the story. Then the Singapore Story will never end', the Prime Minister explained. Under the leadership of Professor Tommy Koh, chairman of the National Heritage Board, the two projects are to be financed by $25 million from the government, with the rest of the total $63 million coming from public donations.[13]

These are fitting gestures to remember the life of the great man. Not that there is any fear that he will be forgotten. The future of the Singapore Story is however less certain; and that, in spite of the Prime Minister's assurance, is the worry. Lee's eightieth birthday, economic recession and the terrorist threat all make people rethink the Singapore Story, to revaluate its messages and political lessons. For the State, however, the end of an epoch may spell dangerously the end of the narrative about life's uncertainty and the ethos of striving for economic betterment. And that would be intolerable because it threatens the collective virtues that have seemingly served Singapore so well.

The problem is of course not unique to Singapore; any final closing of an era will call up the exciting vista of change or the vision of doom with the sweeping away of the old, depending on whom you are talking to.

When French philosopher Alexandre Kojève speaks of the end of history, he describes a world where nothing new happens, where contemporary events are but realignments of the set pattern. The world that he has in mind was created by the aftermaths of the two World Wars, so bloody and radical that they redefined 'the origin of human community' and questioned forever 'the very foundation of the self – of humans... in contradistinction to animals'.[14] As history consists of human endeavours and struggles and achievements that give human life form and meaning, the end of history is the end of these things as well. The outcome is that these activities, evocative of tradition and customary values, no longer help to decide what is morally significant and socially worthwhile. More than that, people begin to 'live according to *formalized* value – that is, values completely empty of all "human" content in the "historical" sense'.[15] The 'post-historical Man must continue to *detach* "form" from "content"'.[16]

In short, the disappearance of 'man proper' means the disappearance of the culture and symbols that define his humanness. The 'post-historical man' returns to a kind of natural state in which the arts and other cultural edifices are like poor imitations of the enterprises of animals:

If Man becomes an animal again, his arts, his loves, and his play must also become purely 'natural' again. Hence it would have to be admitted that after the end of History, men would construct their edifices and works of art as birds build their nests and spiders spin their webs, would perform musical concerts after the fashion of frogs and cicadas, would play like young animals, and would indulge in love like adult beasts.[17]

It may be hard to visualize such a world, but what is easier to grasp is the loss of meaning, the massive alienation from the ideas and wisdom that previously held the world together. The loss and alienation make almost apocalyptic the post-historical world:

'The *definitive annihilation* of Man *properly so-called*' also means the definitive disappearance of human Discourse (*Logos*) in the strict sense. Animals of the species *Homo sapiens* would react by conditioned reflexes to vocal signals or sign 'language,' and thus their so-called 'discourse' would be like what is supposed to be the 'language' of bees. What would disappear, then, is not only Philosophy or the search for discursive Wisdom, but also that Wisdom itself. For in these post-historical animals, there would no longer be any '[discursive] *understanding* of the World and of self.'[18]

However, in a classic Hegelian move, Kojève also sees chaos as leading to order, the good as coming out of the bad. Revolutions are not only about violence and death; they also remove the cobweb of tradition, wiping clean the slate for building a new society and writing new discourses – even if the new regime sometimes turns to the repressions of old. The end of history incites new hope because the old wisdom and reasoning are no longer germane to social and political life. But the rub is that this world of moral and philosophic renewal can also frighteningly be one of social nihilism, or of decadence, if you like. When Kojève writes of 'the post-historical man becoming an animal again', we recall, he has in mind the return to 'pure form', to the chaste but sterile aesthetics in human activities (he mentions the highly ritualized tea ceremony and Noh theatre in Japan as examples). If decadence is the right word here, then the true nature of 'post-historical man' is his self-possessing enjoyment of all sorts without being mindful of their historical origins and social effects. Since appetite is now the rule, the end of history is also the end of the 'absolutes' and 'enduring values' that underpin the social and political order.

Singapore the never-ending Story

We need something of Kojève's brilliant philosophic vision to summon up what is at stake for Singapore on the edge of a new era. The changes are nowhere near as traumatic as the French Revolution or the World Wars, but we must not belittle their social and psychological effects. The National Father's long life is a cause for celebration; but 80 years is by any standard a sign of approaching mortality to be contemplated in the dark night of forbidden thoughts. As hope for a softer, more forgiving PAP rule by younger leaders is accompanied by the fear of an uncertain future, history's finale is at best a mixed blessing. If history cannot be changed, then the ending of the epoch – and of the values that go with it – can be at least arrested to run at a slower pace. What then, in this manner, is the wish that 'the Singapore Story will never end', as the Prime Minister has put it, except the wish for the intransigence of the national values? For those less caught up with the nationalist passion, the Prime Minister's call begins to sound quietly desperate. Quite simply, 'the never-ending Singapore Story' is a fervent re-reaffirmation of the stuff of the PAP narrative: that the values of personal and collective sacrifices, and compliance with the State, are economically and existentially worthwhile. The closing of the State narrative would bring with it the 'post-historical national subject' in the way Kojève has described him. Such a national subject has no place in Singapore. For he would spell the futility of nation-building, and all the endeavours and heroic enterprises that ennoble the nation and its leaders.

The PAP State's attempt to arrest history's finale is not to recast the nation on the fixed course of the past, however. For the PAP State, the doomed scenario of 'Singapore after history' is perhaps shown up in what it sees as happening in the West, which is mired in the post-industrial malaise of communal breakdown, urban crime, long-term unemployment, and social welfare blowout. The remedies to these ills have never been about returning to the old ways – even if the Asian Values discourse has a great deal of that. Nor is it about carving out of some 'glorious past' in order to build a platform for the future. In any event, it is hard put to find a 'golden age' in a trading post of brief colonial rule, and there was even less of it in Singapore before that. As for Asian Values and the short-lived Confucian education in schools, they are not so much nostalgic as practical measures for putting a check on what is seen as the decline of Asian cultural cohesion in the onslaught of Westernization. Without an Angkor Wat or a Great Wall or a Grand Revolution to inspire and consolidate the passion of 'cultural nationalism', the history of Singapore is always the shoddy 'thing' invested to do the ideological work. The past is no loss for Singapore in the manner that it would be for Cambodia or China or France, because there is relatively little that stands in the way of inventing one.

The end of history, one is tempted to say, is very much a double-edged thing for Singapore. A short history and the lack of a Great Civilization make Singapore the poorer by depriving it of splendid 'cultural resources' for all kinds of purposes: to drive nationalist sentiment, to construct a present linked to a long, continuous past, to fashion a nation that has come out of the 'great revolution' and so on. If all these grand, sublime enterprises sound bleakly out of place in Singapore, that is precisely the point. The littleness of Singapore – the littleness of geographical size and population, the littleness of democratic passion and moral imagination – is matched by the shallowness of historical depth and collective memory. Yet the littleness and shallowness are not without their benefits. Singapore might even congratulate itself on them. As it looks at most of its neighbours in the region, Singapore can point to religious divides and deep historical traditions that seemingly put them in the quagmire of cultural backwardness and economic stagnancy. Singapore, the proud inheritor of the British 'shopkeeper' in the East, is luckily free of them. The island nation lives on the knife-edge of the new without the burden of the past, where the fruits of capitalist modernity can be enjoyed without the moral and religious reservations of some of its neighbouring countries. Unlike Malaysia and Indonesia, there is no fundamentalist zeal to denounce its materialist enterprise, to decry the lack of spiritual piety of its pursuits.

Bleakly announcing an indeterminate future, yet promising opportunities unshackled by the past: the Singapore Story is a paradox. In barely five decades, it has gone through transitions from self-government to independence, manufacturing to post-industrialization, a Third World to a First World nation. Each is a breathless passing of one epoch to another. As the new promises greater prosperity and national prestige, there is no need for nostalgia (and serious heritage preservation at the expense of development). What is passed heralds the new and the better which itself quickly vanishes with the coming of the next stage of 'economic development' and fresh social and cultural aspirations. And it is hard for visitors not to feel the infectiousness of all this. They cannot but be struck by the audacious vision that in Singapore anything of the future is achievable: a city of prosperity and culture, of a stock exchange and art centres, a city thriving on creative industries and an innovative economy.

Visitors also find it hard to describe such a world. Without the burden of the past, and where nothing seems to stand in the way of new achievements, Singapore becomes dangerously close to living without 'absolutes', as Kojève might put it. Without 'absolutes' there will be no enduring principles that can ensure some kind of continuity of the past with the present and future. This is the dark realization that has ushered in so many cultural campaigns in recent years. Asian Values, Confucian education, *Shared Values* and the ethics of the community desperately search for universals on which Singapore can build a life of existential meaning and richness. And further still, the hard determination that judicial caning is nothing but just, painful punishment, that only the State can decide what is natural and what is not, we recall, helps to anchor the society on to some essential, unshaken foundation of values. In Singapore as elsewhere, the end of history also drives the elegiac need for certainty, and the search for 'the (true) principles which are at the basis of [man's] understanding of the World and of himself'.[19]

Conclusion (of sorts): the undying National Father

In 1984 the idea was raised about the fear of future governments using the huge national reserve for various purposes. There had to be some way of protecting the fund and discouraging the government from squandering it to appease voters, from spending it in extravagant projects to buy electoral support. The scenario was disastrously opened up by what the PAP circle called 'the structural weaknesses of parliamentary democracy' where voters could through the ballot box exert pressure on the government. Extravagant public spending is tempting for any government in power, by drawing on the national reserve. In the words of the government White Paper:

> In many countries, irresponsible free-spending governments have mismanaged the national finances and irreversibly ruined their economies. When a government sets out to spend money on generous subsidies, dispenses largesse in order to bribe the electorate, it has to do so by raiding the country's financial reserves or by raising large international loans for consumption rather than investment. Before long the country, no matter how rich or well endowed, approaches bankruptcy and economic growth comes to a halt.[20]

The outcome was the elected President, who since August 1993 has become more than ceremonial and has taken on 'veto powers over budget decisions' and over government 'spending from financial reserves'.[21]

When the idea was first mooted, there was much speculation whether Lee Kuan Yew would become the President, as he was preparing to hand over power to the next Prime Minister, Goh Chok Tong, after the 1988 general election. However, Lee told the audience in a public speech: 'I don't have to be president and I am not looking for a job. Please believe me.'[22] Nonetheless, the elected presidency as guardian of the national coffer was very much Lee's brainchild. And it seemed most logical that he should take the office. Not only was he morally and politically the best qualified; there was also the prospect of his retirement after the 1988 election. Of course, as we have seen, retirement is not his option and, besides, then as now he 'need not become president to remain influential'.[23]

All that took place more than ten years ago, but the elected presidency was an unforgettable rehearsal of what was later brought up by his eightieth birthday. If not history's end, then mortality was then as it is now the uneasy subject. When Lee quickly moved to foreclose on the subject of him taking the office of President, his reply was dramatic and with a touch of the ghoulish; the *Straits Times* reported:

> Mr Lee said that as a member of 'that exclusive club of founding members of new countries, first Prime Ministers or Presidents', he could not disengage himself from Singapore. 'Even from my sickbed, even if you are going to lower me into the grave and I feel that something is going wrong, I will get up. Those who believe that [I would not do so] when I have gone into permanent retirement, really should have their heads examined', he said. . .[24]

Perhaps the august National Father too, and not only the citizens, could do with a bit of examining of the head and gentle probing of the unconscious. For Singaporeans, the wish for his retirement is a wish for the freedom from his continuing firm hand in national affairs. However, this is always mixed with a great deal of fear for a world without his political wisdom and powerful influence. Thus the Singaporean subject is caught: Lee's lasting political influence is a national blessing, but it is also something which people must struggle with in order to break free.

Still, the undying National Father remains richly meaningful. Nothing is more suggestive of the wise benevolence of the elderly than their unceasing duty of care – even from the grave. With elementary Freudian insight, we say that a father of eternal life, who refuses the natural logic of mortality, is also a father who refuses to let go. And a father who refuses to let go thwarts the maturity of the young, whose arrival in the world demands the critical battle with – and, as the anthropologist would say, the symbolic slaying of – the father. In the psychic drama of Freud, the son's Oedipal struggle against the father is over the authority and resources – material and sexual – he holds. Maturity is a matter of breaking out of the shadow of the father, whose symbolic demise – we are talking of the symbolic play of the unconscious – allows the son to come into his own. The undying National Father works against all this. A Lee rising from the grave conjures up the image of British actor Christopher Lee in one of the gaudy offerings from the Hammer studio, as the Dracula impersonator opens the coffin to begin his nightly prowl for blood and barely disguised sexual conquest. This image may be only for the less reverent-minded Singaporeans; it nonetheless intimates the fearsome psychological drama of a National Father undead. He would rise from the grave to put things back in order just as he would put a curse on those who betray his ways. We are perhaps making too much of a casual remark made in a public speech. But for Freud slips of the tongue, like jokes and clumsy accidents, are significantly more than what they are; and even people of sobering intellect and public propriety cannot escape them absolutely. At the last, the National Father who would get up from the grave announces his victory in the Oedipal conflict, condemning the son to subservience and as sexually wanting.

This is the language of the unconscious; but it is not too much to see the ambivalent longing for Lee's retirement from the window of psychoanalytic drama. What is this longing except a secret beholding of the father's real, lasting demise, and thus the emergence of the self-determining subjects from the shadow of his benevolent but crippling authority? It is also a crucial insight that few could really dare to contemplate a Singapore without Lee, the prospect of life without PAP rule. In Singapore, so limited is the political vision that speaking of the country without the PAP is like speaking of an independent state of Hong Kong free from Chinese rule, as unintelligible as the Vatican announcing the use of contraceptives for Catholics. The tragedy for Singapore, in the midst of material abundance and modern opportunities, is its narrow political horizon and vision. As people return the PAP government term after term, the self-making Oedipal struggle barely breaks through. What the electoral victories confirm, we might say, is the fearsome prospect of Singapore without PAP rule. Against the uncertainties, Lee has counselled the need of moral fibre and communal values. The rub is that these are not a matter of existential search or private conscience but ones of State enterprise. Making a huge amount of the importance of social peace and political consensus, the PAP State's search for 'fixity', for some enduring values, is a constant anxiety. The book has described the strange and elegiac flowerings of this anxiety, most spectacularly in what I have called the 'totalitarian ambition': the unfulfilled desire to define and shape all human aspirations and all things social and political. 'Totalitarian ambition', even when it cannot be realized in practice, pacifies 'the traumatic' in the national history just as it offers a panacea for the pain of history's passing.

If the National Father is symbolic of all that is constant and good, then he has to live on for ever. His eternal life freezes all the principles that he has bestowed on the nation, and refurbishes the national ethos in danger by the end of history. The National Father of eternal life halts the end of history and, through his continuing, energetic cajoling, arrests the end of human struggle as a meaningful and fertile undertaking for building the future. For Singapore the struggle of the Social Darwinian sort has to go on because it is the impeccable logic of life and economic competition. People may give different evaluations of the worth of this logic; but the essential and enduring truth cannot be contested. This may well be the sublime enterprise of the Singapore State at the end of history: to open the society to the old form and significance time-tested to have wondrously worked, while recognizing the pull of new forces and to make partial adjustment to them. The eternal National Father helps to firm up the project of the PAP State by offering a template, a pattern of ideas. With the ideas he richly symbolizes, shallow of history and ideologically constructed as they are, nothing is at stake except people's willingness to participate in them, to be part of the nation's ever-brighter future on a path of evolving 'new histories'.

9   Epilogue

Useless pragmatism

> We act with certain intention; then our act produces results which we
> did not intend and could not have foreseen. Thus often we reap more
> blame and disrepute, and occasionally more praise and honour than
> we deserve.
>
> Freud, *Woodrow Wilson*

So in these anxious, exuberant ways the Singapore State makes its
appearances and works its magic on the people and the international
world.

Since these are affairs of culture and settled habits, they do not
change radically. Observers are of course wont to look for 'patterns'
in the official policies and their banal, quotidian effects. But the State
is itself deeply committed to keep to the customary approaches: the
Singapore Way should be upheld because 'it works', so the refrain
goes. The task of the commentator is thus made easier. More exactly,
it is not so much that things do not change as that what is new
always retains some essential elements of the old. And these elements
we recognize as the 'foundation' of Singapore: the restless 'culture
of excess' with which the PAP State makes and remakes itself, and
instils its visions and values on the people. The most lasting effect,
as I have made plain, is to give a powerful urgency to all aspects of
life in Singapore, thus binding the State and society to the common
purpose of striving ambitions. The 'culture of excess' may be fretful
and apprehensive; it also gets things done.

Singapore is the most eloquent argument that state-led economic
rationalism works. But then very few liberal-democratic regimes have
Singapore's labour laws and high level of compulsory saving through
the national superannuation scheme, the Central Provident Fund
(CPF), which the State can use to finance infrastructural investments.

Very few too have Singapore State's freedom to manipulate the democratic form so as to allow it to intervene in so many areas of social and economic life. These are the good bits of the Singapore Story: how the State's social policies work cheek by jowl with the economy, transforming the country from a Third World to a First World society. Sometimes you wonder when the Western conservatives look towards Singapore whether they do not secretly long for the smooth economic and labour reforms taking place there. Perhaps their own countries could do with the politically near-painless ways of reining in the trade unions and civil society out to obstruct the government's great purposes. In France, Prime Minister Dominique de Villepin faced street riots and public protest over the new labour law making it easier for employers to fire workers, and Australia's Prime Minister John Howard finds his workplace reform slowly being bogged down by legal and legislative difficulties. Wouldn't they wish for some of the Singapore methods if they could get away with it?

For the Singapore State, however, its harsh social measures are simply 'causes' of the country's economic prosperity. This is as much a justification as an explanation. For if harsh social measures could indeed bring about economic growth, then the former Soviet Union would have been the most prosperous country in the world, and Zimbabwe an economic powerhouse. One wishes it were that simple. Still the evident material accomplishments do tend to silence most criticisms of Singapore's shortcomings. Economic prosperity makes it easy for Western observers to acquit Singapore of the charges of undemocratic practices; when international bodies speak of Singapore's lapses in human rights standards they do not align it with the worst violators in Asia, Africa and Latin America. Besides, these are the times of economic rationalism, mass consumption and transnational capitalism. It is hard to see how Singapore's devotion to materialist ends can be wrong and ethically worthless. And indeed, how can we fault Singapore when similar ends are being pursed with equal keenness by our own government at home?

There is for me something morally insipid about this kind of appraisal. If Singapore's prosperity is indeed brought about by the bleak and joyless policies, then there are costs as well to be paid. 'There is no free lunch' applies as much to the Singapore State as to the philosophically clueless. This is the other side of the Singapore Story: the dark moral vacuity that comes out of the PAP's famous pragmatism, its obsession with 'things that work' and 'the ends justify the means'. And perhaps we should play up to the PAP's own game. For moral discernments too are, in a special sense, practical issues. They have an importance, and an influence over us, that we do not control. In Singapore, there is no escaping the fact that the 'moral performance' of the PAP State defines the political rule, and shapes the quality of social and political life. And 'moral performance' here makes us think not of Asian or Confucian Values, or even the moral authority of the PAP leadership, but of the State's blindness to what its own actions and policies mean in ethical terms. There are surely other issues about the caning of criminals on vandalism charges and a mandatory death sentence for possession of cannabis than law and order. The forms of punishment and the singularity of their meaning speak of the State's supreme confidence that it is right. Where such confidence comes from raises interesting questions about the State's hold on power and its moral imagination.

The education of desire

The Singapore Story is a grand experiment in liberal democracy. The bones of the story are those of talented, visionary men who had a strong sense of history, who, educated in Britain, nonetheless foresaw colonialism's ending and the coming of national independence of the former colonial world. The young Lee Kuan Yew and his colleagues were by personal inclination and education hard-nosed pragmatists, as it has been often said. Tracing his early political career, and sieving through his speeches and pronouncements of the early years, as I have done, we cannot but note of him 'learning on the job', of working out his own ideas in the social and political context of the time. His education did not stop at Cambridge University; for the young lawyer-politician in his twenties, there was much to discover in the encounters and machinations with friends and enemies of all sorts. Much of his pragmatism was learned, and pragmatism was simply the political accent of the time. For liberal-democrats like Lee, the massive problems of unemployment and housing and poverty had to be solved and solved quickly. This was the only way to ensure PAP's electoral success and defeat of the Communists and leftist radicals. For many people, communism was more than armed struggle and jungle warfare; it was first and foremost a means of achieving social and economic security. It is easy in hindsight to ridicule the eerie dullness of the fantasy of 'communist utopia as a land of abundance and equality', but the hope and urgency it evoked was very real. Han Suyin, the novelist and chronicler of Mao's revolution, wrote in 1960:

To divide the world into communist and anti-communist faiths is to obscure realities, not to explain the monstrous necessity which drives men into action. The differences are in speed and method toward a common aim: food, shelter, social security, a living wage, social justice, education.... In Asia today whichever country or nation is going to achieve this basic social security within the next twenty years for the greatest number of its people, is likely to set the pattern for others.[1]

In this volatile environment Lee and his PAP colleagues honed their political skills. They learned the importance of achieving wealth and security for Singapore as the best way of defeating communism and its powerful appeal. Adopting the economic priority was the important political wisdom for anyone aspiring to power and leadership. Some in the trade unions amended their radicalism and came into the PAP fold, and the party leaders learned to weed out the excessive socialist zeal. Pragmatism quite naturally came to galvanize the political thinking: that economic security was the highest priority, and moral self-scrutiny an ill-affordable luxury.

I like this version of the PAP's famous pragmatism. The anthropologist Ann Laura Stoler, writing about Dutchmen in the colonial East Indies, describes how they were drawn to the irresistible attractiveness of the native women, and were keenly aware of the European hygienic theory about the importance of discharging pent-up sexual energy.[2] Men's desire was a matter of the hard prompting of lust and European health measures; Stoler calls the process 'education of desire'. Shorn of the language of sex, 'education of desire' could easily be adapted to describe the formation of PAP's driving ethos. The inner wishes of the PAP men, as of ours, do not spring mysteriously from the recess of psychology; they are also to a large degree acquired and related to the contingency of time and circumstance. PAP's pragmatism can be shown to have a social origin and political logic – even if it is full of clinching paradoxes.

In hindsight, nation-building in Singapore has shown that, faced with the British withdrawal and the massive social and economic problems, it did not go the way of dictatorship and political corruptions that seemingly plagued some states in postcolonial Asia and Africa. This is a credit to Lee Kuan Yew and the party he brilliantly led. It is a tired political saying that revolution is hard, and harder still the task of reconstruction after the revolution. In this task the quality of leadership, its ability to win over the masses to the national cause, and the nation's geopolitical position that invariably defined friends and enemies hugely mattered. These particularities aside, democracy also presented a more general problem. In regard to Singapore, we recall, it is the elegiac vision of Lefort[3] that helps us to open up the shortcomings of democracy in the uncertain condition of the early years of self-rule and nationhood.

Democracy disperses power to 'everyone', and electoral politics is often intolerant of strong leadership and policies that demand sacrifices from the voters. Politics in a democratic society demands at best a series of compromises: between leadership and popular assent, collective good and sectional interests, and short- and long-term needs. Given the social and economic problems of the 1960s and 1970s, an argument can be made that the PAP State had legitimacy on its side when it introduced many drastic measures to ensure Singapore's survival. Pragmatism was the wise and necessary approach. It narrowed down the priorities to the urgently economic, trimming the policy choices so that resources and energy could be best directed.

The tragedy of the PAP State is how it continues to hold on to an outmoded model of economic practicality in a different, more peaceful time. In the process it is in danger of slowly frittering away the moral credit that used to provide legitimacy and popular support.

Vignette one

I last went back to Singapore in March 2006. The invitation came from the Singapore Management University (SMU) to attend a conference in its downtown campus in Bras Brasah Road. The Third International Chinese Entrepreneurship and Asian Business Networks Research Conference on Value Creation through Knowledge Governance, to give the full title, dealt with knowledge management (KM) in business corporations and attracted academics from China, Switzerland, Germany, Australia, the United States, Great Britain and Singapore. The conference piggybacks the Chinese business ways on the newer, sexier subject currently much in vogue in official circles. KM reworks the idea of corporate culture and puts emphasis on innovation and communication. Terms like 'collective intelligence', 'smart companies' and 'knowledge repositories' cry out for creativity and information-sharing in corporations in these globalizing times. I listened to the papers, and the breathless reminder of the professor from Basel University; 'knowledge is a factor of production in the high-tech services economy', he said. Since KM calls for, and brings about, innovative thinking and smart, creative people, the Singapore government is going full steam ahead with training the civil service on KM. One SMU lecturer told the conference that he has advised Singapore Airlines and a government ministry on how to improve creativity and customer feedback. Singapore's interest in KM is no secret. The morning papers are full of advertisements from universities and colleges offering year-long courses and six-month diplomas in innovation, entrepreneurial expertise and business leadership. A degree in 'leadership and innovation' has replaced the MBA as the hottest item for bright young executives.

One has reason to be impressed. Here is another example of the State leading the people and taking the nation in the new direction of the knowledge-based economy. Perhaps trying to inject a bit of cool-headedness in all this euphoria, I put the questions to the professor from Basel University: Can innovation and leadership be taught, in the same way computer science or accountancy can be taught, in the structured, objective-leaning mode of the university classroom? Don't creativity and innovation require an environment of risk-taking and bold, far-sighted individualism? Perhaps teaching creativity is a contradiction in terms, I meekly suggested as I sat down.

The professor answered 'I know where you are coming from', and turned to the more important questions from the floor about how to increase company values through KM and how to measure KM performance. Still, it was not hard to know that one was on the right track. The Defence Minister Teo Chee Hean, a bright star of the new generation of PAP leaders, told the press that he was setting out the role of the civil service in the new era:

> He named three fundamental weaknesses: they were too risk-averse, they lack understanding of how the market worked, and they didn't cooperate enough with other ministries.
>
> From now on, public employees...would be appraised by new ideas they come up with.... Keeping quiet, simply following orders and focusing only on implementation would not be a career-enhancing strategy.[4]

And the news report gave the explanation:

> The problem is not entirely the fault of the civil servants. They take orders from the minister in charge. Some of their problems lie in a political leadership that is generally reluctant to deregulate too quickly for fear of things will go wrong.[5]

One hesitates to belabour the point. The days of union busting and detention of radicals and Marxist conspirators are gone (though the laws that enable the State to do so are still on the books), and the pursuit of innovation and entrepreneurial leadership seems to present a new yet familiar sense of urgency. The peevish, self-serving compliance of civil servants is not exactly wartime-like crisis, but it nonetheless brings out the old disquiet, and with that the inflated confidence – that Singapore can 'manage' itself out of the problem. The idea that talent and free-thinking can be 'bought' with fee-paying university courses, that civil servants can become more creative by their taking up management training: it sounds like the pragmatism of old. Yet there is nothing straightforward about the idea. If things are evaluated in terms of their usefulness and how they bring beneficial outcomes, then the thinking that the means justifies the end is a powerful lure. When attention is obsessively given over to 'things that work', there is no need for moral curiosity and self-examination. The result is an ethical blindness, one that muddies the PAP hold on power, giving it a certain hard-to-ignore murkiness.

Vignette two

In 2006 it was also the year of Singapore's eleventh general election. The PAP won and retained 82 of the 84 seats. The victory was so total as to mock the idea of 'parliamentary majority'. Even assured of spectacular winning, the PAP nevertheless fought the contest with an iron-fisted seriousness. The two opposition wards were like so many gnats landing on the skin of the PAP pride. Potong Pasir has been held by the Singapore Democratic Alliance for the past 25 years,[6] and Hougang by the Workers' Party since 1991. Every effort was made to win back the two seats. The government warned that the constituencies voting for the opposition would not be included in the state-funded upgrading scheme. Without upgrading there would be no improvement of the parks and playgrounds, and better facilities for the elderly such as ramps and lifts that stop on every floor. In short, as one minister put it, those housing estates with an opposition MP risked becoming 'slums'.[7]

For residents of Potong Pasir and Hougang this was nothing new; they had heard the same government threat in the 1997 poll but voted in the opposition nonetheless. They are public housing 'heartlanders' who are at the lower end of the economic gap and most vulnerable to the economic downturn. The protest votes vented some of their frustrations. In the past the 'punishment' of opposition wards did not receive much of a public outcry. The majority who voted for the PAP did not suffer, so it wasn't their concern. This time however the

PAP move seemed to open up a good deal of unhappiness. Letters to the papers were full of outcries like 'nauseating', 'Mafia election' and 'unbecoming trick'. The general feeling is that upgrading uses public money and should be enjoyed equally by all Singaporeans; it should not be used to blackmail voters. One reader wrote, 'If people give in to this, it will be a sad day for Singapore. It is downright low in characters [*sic*] that our Ministers had to resort to this to win votes.'[8]

With the crown of history on its head and after four decades of brilliant rule, why did the PAP resort to what even its own supporters would see as cheap political antics? For the ruling party the election is not about winning – on that it has no doubt – but the margin of electoral victory. Given the feeble opposition, it is the margin together with voters' turnout that is the indicator of government popularity. A big margin of electoral win, in the PAP's view, is the source of its 'moral authority to rule'. With such things, you simply do not put a figure to them. Spectacular wins and a parliamentary majority are like the digits in the balance of one's bank account: at what point do you say you have enough? Besides, one has to show senior leaders what one is made of, that one has the same ruthless determination to win as they had in the past. The new PAP team has inherited not only the easy electoral victory, but also the method – the strategy and assumptions – of winning. If they had questioned the punishment of opposition wards in party meetings, the public did not know about it. In winning results matter, they must have learned.

So punishing the opposition wards is simply pragmatism, and what is pragmatism except equating truth and justice with usefulness? Still one wonders: were the party strategists who planned the election tactics so oblivious to the ethical implications? One can only guess that they probably did not care, or that they knew they could get away with it. After all the party again won the election.[9] The public uproar undoubtedly forced the party to some rethinking, but the people saw no signs of soul-searching and excruciating self-doubt. Business was, by and large, as usual.

The 2006 election was the first time Lee Hsien Loong had had to face the electorate since he became Prime Minister two years previously. Leading the government in its first general election, he had much to prove. He told the media conference on 8 May, two days after the election, 'We have a lot to do and we are starting now. We have a new leadership team in place; it will see Singapore through the next 15 to 20 years.'[10] It is almost churlish to say that Singapore is among the few countries in the world where the ruling party can confidently announce that it will still be in government 15 to 20 years on. But there is no mistaking that the Prime Minister was anxious to get on with the job. And with the strong mandate, he will let nothing stand in the way of carrying it out. 'Suppose you have 10, 15, 20 opposition members in Parliament. Instead of spending my time thinking what is the right policy for Singapore, I'm going to spend my time thinking ways to fix them, to buy my supporters' vote', he said.[11] The Prime Minister later apologized. He explained that what he meant was that he would have to spend time fighting the opposition instead of attending to the issues of the state. Still the point was memorably made.

Power and blindness

Something can be made of what is evidently a slip of the tongue. The Prime Minister did not mean what he said: this is a reasonable enough defence as defences go. But a slip of the tongue is also a habit, an intuitive running through the old ways. It is less a mistake than the mind giving over to the 'truth' and the irrepressible urge to say it, and neither political protocol nor the rules of parliamentary democracy held much sway. The opposition as a kind of inconvenience, to be tolerated in form, when there is the business of governing to attend to: this is pragmatism too. The opposition is not 'useful' and does not bring happiness and prosperity to Singapore, so the PAP might argue. Here usefulness suddenly reveals its lack of moral dimensions. When you measure truth and the ethical with the yardstick of usefulness, the outcome is moral vacuousness.

It is hard to avoid the conclusion that in Singapore pragmatism is political ideology par excellence. As political ideology, it conceals and mystifies. Pragmatism, buoyantly nestling at the centre of political power, is promoted as the only viable vision for the nation. Since practical state policies have brought visible abundance to Singapore, the moral ambiguities of the State's actions can be explained and justified. And inevitably, Singapore's success story also blinds the State to its moral defects.

It is easy to show that, as a political ideology, the PAP pragmatism is less consistent and less endowed with practical sense than it has been normally presented. The cases I have examined in this book seem like easy targets simply because they are hard put to stand up to more thoughtful and rigorous critique. Take just one more example. In Singapore, government ministers are paid huge remunerations commensurate with the wages offered by the private sector.[12] In 2003 the Prime Minister's annual salary was $1.74 million, or US$980,000, more than twice the US President's US$400,000. Junior ministers are paid about $870,000 a year.[13] The argument is that high remuneration is necessary to attract talented people into the parliament and, given high wages, they are less likely to be corrupt and accept bribes. It is now a different time, and the 'period of revolutionary change that threw up people with deep conviction and overpowering motivation is over', Lee Kuan Yew said to put forward the case.[14] For a State that cares so much about 'values', it seems remarkably lacking in moral magnitude. People are asked to believe that, in the final analysis, people who enter the parliament keep their original ambitions in running a shipping line or a bank or a retail giant. And perhaps there are still people who seek politics as a vocation, people motivated by personal sacrifice and duty to serve above money and professional mobility; some of these people might even be found among the political opposition. But this is not a line of thinking that can be pursued in Singapore.

On the other hand, the State can be easily taken to task because it is uncompromising about its principles of political rule. The principles may be to some degree put to the test during elections, but party leaders abhor what they see as blackmailing by voters and public opinion. Sticking to the determined ways and being unswayed by the changing wind of electoral favours are considered the primary virtue of political leadership. This self-absorption defines a great deal the nature of the PAP's hold on power. Everything about the PAP – its restless desire, brilliant administrative skill, and self-confidence – seems to make sense this way. No one need quarrel with the on-the-spot fines for people caught with dengue larvae breeding in their garden-pots, or the high registration fee and tariff duty paid for private cars in a small, densely populated island. Discounting the general unhappiness over these measures is right and appropriate; they tell of the State's concern for the public good. But other practices are potent of obsessive self-regard blind to ordinary reasoning. How else are we to make sense of the aggressive denigration of the political opposition when in the United Kingdom it has the honourable title of Her Majesty's Opposition? Or the shabby reasoning of the ruling on fellatio that only the government can decide what is natural and what it not?

Self-absorption is also an arrogation of power.

When pragmatism operates as ideology, it slashes and burns what it identifies as unrealistic thinking and sentimental liberal soft-headedness. In the PAP State's scheme of things, usefulness is undoubtedly crucial; so is the semantics of it. Pragmatism squeezes the last drop of significance out of Singapore's prosperity and efficient government: it is less about the practical sense than for propping up an important idea – that even ethics and justice can be decided by the broadly considered usefulness. The old adage about the corruption of power augurs the tendency of political power to look inwards and to answer to itself because it knows what is best for the people, because the leaders are persons of moral distinction. The arrogation of power is the most potent when power is blind not only to outside disapproval – but also to itself.

Power and the tragic

Despite the glib jingoism, the transformation of Singapore into a First World society is no small feat by any standard. There are however moral costs to this. With the laurel of achievement on their heads, the State leaders can install themselves on the Olympic height of authority and eminence. There is little need for moral self-scrutiny, and people are asked to accept their evaluation of things. It is the Greek philosophers who reminded us of the tragedy of Great Men. Endowed with moral excellence and social distinction, Great Men are also prone to the universal working of the 'tragic flaw'. For Aristotle, the moving effect of the tragic lies in the secret vulnerability which such men are not aware of and helpless to change.[15] You may be a king or a prince or, in our parlance, a rich nation, but fate or destiny imposes its logic and turns you around. For all their ambitions and strivings, men and nations are made little and insignificant by the larger design of things. It is this notion of the tragic, I have always felt, that helps to express the sense of wonder and regret when I think of Singapore. Here is a nation that has fought the big fight against colonialism and radical violence, and arguably some of the excesses of its measures could be excused by the circumstances of the time and exonerated by the peace and prosperity they brought. As for the PAP leaders, with their sophisticated and realistic evaluations of things, why are they filled with such gloomy predilections for imagining the worst for Singapore? The days of political heat and massive poverty have long gone. If not personal freedom, then the new knowledge-based economy demands the loosening up of the State's tight control. The old pragmatic way has passed its 'use-by date'. Whatever the State's own reasoning, instinctively sticking to the customary ways is a telltale sign of power's blindness and presumption. The brilliant people in the PAP have a gift for administration and quick and decisive action, but this is not quite the same as a gift for moral imagination. To say this is not to wish the country ill. Singapore is still, after SARS, the terrorist threat and the economic down of 2001–02, a prosperous place. In 2005 the unemployment rate dropped to 3.2 per cent from 3.4 per cent the previous year.[16] The economy has also recovered. Growth for

2005 was registered at 6.4 per cent, though the figure for 2006 will probably be less, at 5.9 per cent, according to government forecasts.[17] Yet some Singaporeans are beginning to ask: what will the utopia of material prosperity and political consensus look like in the future, say two decades on? The young, well-travelled and spoiled by the freedom of the internet and overseas education, might even ask the bold question: would I want to live and have a family in such a utopia? There are countries freer and even more prosperous than Singapore. For the educated young professionals, migration is a choice. Things of course have not come to this. There is no exodus or massive seething discontent, but doubts about the PAP way and electoral protest, as we have seen, are evidently here. In the end, Singapore suggests a powerful moral lesson. It has brilliantly demonstrated that a new government can free itself from the shadow of European colonialism and be clean and committed to the prosperity of the nation, and now, in the new circumstances, it takes greater imagination to see that national happiness can be achieved differently – by generosity, moral imagination and flexibility of purpose.

Notes

1 The magic of the Singapore State

1 M. Barr, *Lee Kuan Yew: The Beliefs behind the Man*, Richmond: Curzon, 2000a, p. 75.
2 It was also to be an 'anti-Communist' strategy. Lee realized that after a merger the Federal authorities in Kuala Lumpur would suppress the leftists, especially the Barisan Socialis which posed the greatest threat to the electoral success of the PAP.
3 Barr, *Lee Kuan Yew*, p. 75.
4 J. Minchin, *No Man Is an Island: A Study of Singapore's Lee Kuan Yew*, North Sydney: Allen & Unwin, 1986, p. 156.
5 Lee, K. Y., *The Singapore Story: Memoirs of Lee Kuan Yew*, Singapore: Prentice Hall, 1998, p. 16.
6 Minchin, *No Man Is an Island*, p. 156.
7 Ibid.
8 Lee, K.Y., *From Third World to First: The Singapore Story: 1965–2000*, Singapore: Singapore Press Holdings and Times Editions, 2000, p. 25.
9 Ibid.
10 Lee to Malaysian students in London, 10 September 1964, quoted in Barr, *Lee Kuan Yew*, p. 75.
11 Singapore gained independence from the British by being admitted into the Federation of Malaysia in 1963; it became a nation-state after separation from the Federation on 9 August 1965.
12 Lee to a symposium organized by the Historical Society of the University of Malaya, 28 August 1964, quoted in Barr, *Lee Kuan Yew*, p. 76.
13 Lee to Malaysian students in London, 10 September 1964, quoted in Barr, ibid.
14 See note 11.
15 The policies of communitarianism are described by Chua B. H., *Communitarian Ideology and Democracy in Singapore*, London; New York: Routledge, 1995.
16 M. Taussig, '*Maleficium*: State Fetishism', in E. S. Apter and W. Pietz (eds), *Fetishism as Cultural Discourse*, Ithaca, NY: Cornell University Press, 1993, p. 218.
17 Ibid., p. 217.

18 P. Abrams, 'Notes on the Difficulty of Studying the State', *Journal of Historical Sociology* 1, 1988, p. 82.

19 Ibid., p. 79; emphases in original.

20 Q. Skinner, *The Foundation of Modern Political Thought*, Vol. 2, Cambridge: Cambridge University Press, 1978, p. 353.

21 A. Vincent, *Theories of the State*, Oxford: Blackwell, 1987, p. 10.

22 D. Held, J. Anderson, B. Gieben, S. Hall, L. Harris, P. Lewis, N. Parker and B. Turok (eds), *States and Societies*, Oxford: Robertson in association with the Open University, 1983, p. 1.

23 Thomas Hobbes, 'Leviathan', in D. Held, J. Anderson, B. Gieben, S. Hall, L. Harris, P. Lewis, N. Parker and B. Turok (eds), *States and Societies*, Oxford: Robertson in association with the Open University, 1983, p. 68.

24 Ibid., p. 71.

25 Vincent, *Theories of the State*, p. 20.

26 J. Locke, *Two Treatises of Government*, ed. P. Laslette, Cambridge: Cambridge University Press, 1963, p. 309.

27 E. Durkheim, 'The Science of Morality', in A. Giddens (ed.), *Emile Durkheim: Selected Writings*, Cambridge: Cambridge University Press, 1972, p. 98.

28 Ibid., pp. 98–99.

29 *Straits Times*, 20 November 1994.

30 *Straits Times*, 4 December 1994.

31 F. T. Seow, *To Catch a Tartar: A Dissident in Lee Kuan Yew's Prison*, New Haven, Conn.: Yale University Southeast Asia Studies, 1994.

32 Chua, B. H., '"Asian-Values" Discourse and the Resurrection of the Social', *Positions: East Asia Cultures and Critique* 7 (1), 1999, pp. 573–592.

33 Government of Singapore, Ministry of Information, Communications and the Arts Press Release, 11 May 2002.

34 *Straits Times*, 17 June 1994.

35 Chua, '"Asian-Values" Discourse', p. 582.

36 K. S. Goh, *The Practice of Economic Growth*, Singapore: Federal Publications, 1995, p. 104; emphases in original.

37 P. Smyth, *The Economic State and the Welfare State: Australia and Singapore 1955–1975*, Singapore: Centre for Advanced Studies, National University of Singapore, 2000, p. 7.

38 L. B. Krause, 'Government as Entrepreneur', in K. S. Sandhu and P. Wheatley (eds), *Management of Success: The Moulding of Modern Singapore*, Singapore: Institute of Southeast Asian Studies, 1989, p. 443.

39 Title of a collection of essays edited by Devan Nair; see C. V. D. Nair, *Socialism that Works: The Singapore Way*, Singapore: Federal Publications, 1976. Nair was then the chairman of the National Trade Union Council, and became President of Singapore in 1981 but was dismissed from the office in 1985. He died in 2005 while living in Canada.

40 Nair, *Socialism that Works*, p. 188.

41 Durkheim, 'The Science of Morality', p. 99.

42 Government of Singapore, *Shared Values*, cmd1, 1991, p. 10.

43 Durkheim, 'The Science of Morality', p. 101.

44 *South Chinese Morning Post*, 4 October 2002.
45 *Straits Times*, 9 August 2003.
46 Ibid.
47 *Straits Times*, 23 August 2003.
48 S. Freud, *The Freud Reader*, ed. P. Gay, London: Vintage, 1995, p. 598.
49 Ibid.

2 Trauma and the 'culture of excess'

1 *Straits Times*, 12 December 1994.
2 F. T. Seow, *The Media Enthralled: Singapore Revisited*, Boulder, Colo.: Lynne Rienner Publishers, 1998, p. 175.
3 *Straits Times*, 14 August 1993.
4 Seow, *The Media Enthralled*, p. 175.
5 Lee, K. Y., *The Singapore Story: Memoirs of Lee Kuan Yew*, Singapore: Prentice Hall, 1998.
6 W. Benjamin, 'The Storyteller: Reflection on the Works of Nikolai Leskov', in H. Arendt (ed.), *Illuminations*, New York: Schocken Books, 1969b, p. 86.
7 Ibid., p. 87.
8 Ibid., p. 100.
9 Ibid., p. 90.
10 C. Caruth, *Trauma: Explorations in Memory*, Baltimore, Md.: Johns Hopkins University Press, 1995, p. 5.
11 M. Taussig, '*Maleficium*: State Fetishism', in E. S. Apter and W. Pietz (eds), *Fetishism as Cultural Discourse*, Ithaca, NY: Cornell University Press, 1993, p. 218.
12 T. N. Harper, *The End of Empire and the Making of Malaya*, Cambridge: Cambridge University Press, 1998, p. 57.
13 Ibid., pp. 57–58.
14 Ibid., p. 67.
15 T. N. Harper, 'Lim Chin Siong and the "Singapore Story"', in J. Q. Tan and K. S. Jomo (eds), *Comet in Our Sky: Lim Chin Siong in History*, Kuala Lumpur: INSAN, 2001, p. 11.
16 M. Barr, *Lee Kuan Yew: The Beliefs behind the Man*, Richmond: Curzon, 2000a, p. 58.
17 F. K. Han, S. Tan and W. Fernandez, *Lee Kuan Yew: The Man and his Ideas*, Singapore: Singapore Press Holdings and Times Editions, 1998, p. 45.
18 Ibid., p. 257.
19 For the career and sad decline of Lim Chin Siong, see Harper, 'Lim Chin Siong and the "Singapore Story"', and Wee, C. J. W.-L., 'The Vanquished: Lim Chin Siong and a Progressivist National Narrative', in P. E. Lam and K. Tan (eds), *Lee's Lieutenants: Singapore's Old Guard*, St Leonards, NSW: Allen & Unwin, 1999.
20 J. Drysdale, *Singapore: Struggle for Success*, North Sydney: George Allen & Unwin, 1984, p. 172.
21 J. Yeoh, *To Tame a Tiger: The Singapore Story*, Singapore: Wiz-Biz, 1995.
22 M. Chew, *Leaders of Singapore*, Singapore: Resource Press, 1996, p. 79; emphasis in original.

23 Drysdale, *Singapore*, p. 106.
24 Ibid., p. 108.
25 Ibid., p. 109.
26 Ibid., p. 111.
27 Caruth, *Trauma*, p. 4.
28 Ibid., p. 5; emphasis in original.
29 Ibid.
30 Ibid., p. 6.
31 Ibid.
32 C. Lefort, *The Political Forms of Modern Society: Bureaucracy, Democracy, Totalitarianism*, ed. J. B. Thompson, Cambridge: Polity Press, 1986.
33 L. Hunt, *The Family Romance of the French Revolution*, Berkeley, Calif.: University of California Press, 1992.
34 J. Hell, *Post-Fascist Fantasies: Psychoanalysis, History, and the Literature of East Germany*, Durham, NC: Duke University Press, 1997, p. 28.
35 Lefort, *The Political Forms of Modern Society*, pp. 303–304.
36 Ibid., p. 75.
37 Ibid., p. 16.
38 Ibid., p. 284.
39 Ibid., pp. 27–28.

3 'Yellow culture', white peril

1 Lee, K. Y., *The Singapore Story: Memoirs of Lee Kuan Yew*, Singapore: Prentice Hall, 1998.
2 Ibid., p. 332.
3 F. K. Han, S. Tan and W. Fernandez, *Lee Kuan Yew: The Man and his Ideas*, Singapore: Singapore Press Holdings and Times Editions, 1998, p. 139.
4 M. Taussig, '*Maleficium*: State Fetishism,' in E. S. Apter and W. Pietz (eds), *Fetishism as Cultural Discourse*, Ithaca, NY: Cornell University Press, 1993, p. 218.
5 Lee, *The Singapore Story*, p. 326.
6 T. N. Harper, *The End of Empire and the Making of Malaya*, Cambridge: Cambridge University Press, 1998, p. 274.
7 Ibid., p. 292.
8 Ibid., p. 294.
9 Lee, *The Singapore Story*, p. 326.
10 Sai, S. M. and Huang, J., 'The "Chinese-Educated" Political Vanguards: Ong Pang Boon, Lee Khoon Choy and Jek Yeun Thong', in P. E. Lam and K. Tan (eds), *Lee's Lieutenants: Singapore's Old Guard*, St Leonards, NSW: Allen & Unwin, 1999, p. 132.
11 M. Thorpe, '"Penetration by Invitation" – A Lecturer at Nanyang – A Memoir', *Economic and Political Weekly*, 1–8 March 1997, p. 486.
12 Ibid.
13 Ibid., pp. 486–487.
14 Ibid., p. 487.
15 Ibid., p. 435.
16 Lee Kuan Yew, Prime Minister's Speech etc., Government of Singapore Prime Minister's Office, 14 June 1992, quoted in M. Barr, 'Lee Kuan

Yew and the "Asian Values" Debate', *Asian Studies Review* 24, 2000b, p. 316.

17 Since 1962, the University of Singapore, and 1980, the National University of Singapore.

18 D. J. Enright, *Memoirs of a Mendicant Professor*, London: Chatto & Windus, 1969, p. 124; the full text of the lecture can be found in his *Conspirators and Poets*, London: Chatto & Windus, 1966, pp. 48–67.

19 *Straits Times*, 18 November 1960.

20 Enright, *Memoirs of a Mendicant Professor*, p. 127.

21 After writing a conciliatory letter to the government, he remained in Singapore for the next ten years. He returned to England in 1970. A famous poet and critic, he died in January 2003.

22 S. Freud, *The Interpretation of Dreams*, Harmondsworth: Penguin, 1976a.

23 Ibid., p. 389, quoted in S. Pile, 'Freud, Dreams and Imaginative Geographies', in A. Elliott (ed.), *Freud 2000*, Carlton: Melbourne University Press, 1998, p. 210.

24 Lee, *The Singapore Story*, p. 23

25 Ibid., p. 652.

26 Lee, K. Y., *Malaysia: Age of Revolution*, Singapore: Ministry of Culture, 1965, p. 14.

27 Lee, K. Y., *From Third World to First: The Singapore Story: 1965–2000*, Singapore: Singapore Press Holdings and Times Editions, 2000, p. 31.

28 Ibid.

29 E. W. Said, *Orientalism*, Harmondsworth: Peregrine, 1978, p. 12.

30 M. Barr, *Lee Kuan Yew: The Beliefs behind the Man*, Richmond: Curzon, 2000a, p. 310.

31 *Straits Times*, 14 December 1993; all subsequent quotations refer to the same page.

32 From endorsements of Lee, *The Singapore Story*, on the back cover.

33 Ibid.

34 B. Anderson, 'From Miracle to Crash', *London Review of Books* 20 (8), 1998, p. 17; the other factor is Japanese foreign investment.

35 Chua B. H., *Communitarian Ideology and Democracy in Singapore*, London; New York: Routledge, 1995, p. 27.

36 S. Yao, *Confucian Capitalism: Discourse, Practice and the Myth of Chinese Enterprise*, London: RoutledgeCurzon, 2002.

4 Pain, words, violence: the caning of Michael Fay

1 Lim, M. F. C., 'An Appeal to the Use of the Rod Sparingly: A Dispassionate Analysis of the Use of Caning in Singapore', *Singapore Law Review* 15, 1994, p. 23.

2 Ibid., p. 26 and Annex 1.

3 *Straits Times*, 1 May 1994.

4 Ibid.

5 Ibid.

6 E. Scarry, *The Body in Pain: The Making and Unmaking of the World*, New York: Oxford University Press, 1985, p. 4.

7 A. Kleinman, D. Veena and M. Lock, *Social Suffering*, Berkeley, Calif.: University of California Press, 1997.
8 R. M. Cover, 'Violence and the Word', *Yale Law Journal* 95, 1986, p. 1601; emphasis added.
9 *Asiaweek*, 25 May 1994.
10 The most detailed report is given by *Far East Economic Review*, 28 April 1994, from which the events leading to Fay's caning described earlier are taken.
11 *New Straits Times*, 21 April 1994.
12 This and the following media comments, unless otherwise stated, are from *Straits Times*, 8 April 1994.
13 G. Baratham, *The Caning of Michael Fay*, Singapore: KRP Publishers, 1994, p. 32.
14 W. Benjamin, 'Critique of Violence', in *One-way Street and Other Writings*, London: New Left Books, 1979a.
15 Ibid., p. 149.
16 Ibid.
17 Cover, 'Violence and the Word', p. 1602, footnote 2.
18 Ibid.
19 Ibid., p. 1608.
20 Ibid., p. 1611; emphasis added.
21 S. F. Moore, *Law as Process: An Anthropological Approach*, London; Boston, Mass.: Routledge & Kegan Paul, 1978, p. 215.
22 Quoted in A. Josey, *Lee Kuan Yew: The Struggle for Singapore*, Sydney: Angus & Robertson, 1974, p. 129.
23 *Straits Times*, 3 April 1994.
24 *Straits Times*, 16 April 1994.
25 *Asiaweek*, 25 May 1994.
26 Ibid.
27 *Star*, 7 March 1994.
28 F. Bahrampour, 'The Caning of Michael Fay: Can Singapore's Punishment Withstand the Scrutiny of International Law?', *American University Journal of International Law and Policy* 10 (3), 1995, p. 1105.
29 G. Simmel, 'The Stranger', in *On Individuality and Social Form*, translated by K. H. Wolff, Chicago, Ill.: Chicago University Press, 1971, p. 143.
30 Z. Bauman, 'Modernity and Ambivalence', in M. Featherstone (ed.), *Global Culture: Nationalism, Globalization and Modernity*, London: Sage, 1990, p. 143.
31 *Straits Times*, 16 April 1994.
32 Scarry, *The Body in Pain*, pp. 3–4.
33 Ibid., p. 4.
34 Ibid., p. 29.
35 J. Lacan, *Ecrites*, translated by A. Sheridan, New York: W. W. Norton, 1977, p. 4.
36 Ibid.
37 http://web.amnesty.org/library/index/engasa360012004; accessed 20 January 2004.
38 *Star*, 1 February 2004.

39 *Globe and Mail*, Canada, 19 January 2004. See also Comments by Singapore Government, Ministry of Home Affairs Press Release, 16 January 2004, Singapore.

5 Oral sex, natural sex and national enjoyment

1 *Straits Times*, 1 April 1995.
2 Ibid. I am grateful to Joseph Justin Tan for explaining the intricacies of the law on unnatural sex to me.
3 Koh, K. L., 'Unnatural Offences: A Widening Scope?', *Review of Judicial and Legal Reform in Singapore between 1990 and 1995*, Singapore: Butterworth Asia, 1996, p. 364.
4 Ibid., p. 363.
5 *Straits Times*, 29 April 1995.
6 Ibid.
7 G. Heng and D. Janadas, 'State Fatherhood: The Politics of Nationalism, Sexuality, and Race in Singapore', in M. G. Peletz and A. Ong (eds), *Bewitching Women, Pious Men: Gender and Body Politics in Southeast Asia*, Berkeley, Calif.: University of California Press, 1995, p. 197.
8 Ibid., p. 199.
9 Government of Singapore, Social Development Unit, *Annual Report*, 2002.
10 Singapore nonetheless has famously tough drug laws. Anyone over the age of 18 found in possession of more than 15 grams of heroin, 30 grams of morphine or cocaine, or 500 grams of cannabis, is presumed to be trafficking, and faces a mandatory death sentence. See Amnesty International Report, *Action Appeal on Singapore Drug Case from Amnesty International*, 10 October 2003, http://stopthedrugwar.org/chronicle/306/mourthi.shtml, accessed 17 August 2004.
11 M. Foucault, *The History of Sexuality: An Introduction*, Harmondsworth: Peregrine, 1984, p. 6.
12 Ibid., p. 11.
13 L. Trilling, 'Beyond Culture: Essays on Literature and Learning', in L. Trilling (ed.), *The Works of Lionel Trilling*, New York: Harcourt Brace Jovanovich, 1978, p. 51.
14 Ibid., p. 52.
15 F. Jameson, 'Pleasure: A Political Issue', in Formations Collective (ed.), *Formations of Pleasure*, London: Routledge & Kegan Paul, 1983, p. 1.
16 Koh, 'Unnatural Offences', p. 363.
17 Ibid., pp. 364–365.
18 *Straits Times*, 20 November 1996.
19 Ibid.
20 Ibid.
21 Readers may find it useful to go back to the details described in Chapter 2.
22 P. Smyth, *The Economic State and the Welfare State: Australia and Singapore 1955–1975*, Singapore: Centre for Advanced Studies, National University of Singapore, 2000.
23 Ibid., p. 10.

24 A. Josey, *Lee Kuan Yew: The Struggle for Singapore*, Sydney: Angus & Robertson, 1974, p. 3.
25 *Straits Times*, 13 June 2002.
26 W. Benjamin, 'Breakfast Room', in *One-way Street and Other Writings*, London: New Left Books, 1979b, p. 45.
27 Ibid.
28 Ibid., p. 46.
29 *Sydney Morning Herald*, 7 January 2004.
30 *BBC News Online*, 6 September 2003.

6 'Talking cock': food and the art of lying

1 R. Hingley, *The Russian Mind*, London: Bodley Head, 1978, pp. 77–88.
2 Ibid.
3 F. Dostoyevsky, *Diary of a Writer*, Vol. 1, New York: Charles Scribner's Sons, 1949, p. 138.
4 Ibid., p. 133.
5 Ibid.
6 Ibid., p. 138.
7 Hingley, *The Russian Mind*, p. 79.
8 M. Blanchot, 'Everyday Speech', *Yale French Studies* 73, 1987, pp. 12–20.
9 G. M. Joseph and D. Nugent, *Everyday Forms of State Formation*, Durham, NC: Duke University Press, 1994.
10 Talkingcock.com, *The Coxford Singlish Dictionary*, Singapore: Angsana Books, 2002.
11 P. Abrams, 'Notes on the Difficulty of Studying the State', *Journal of Historical Sociology* 1, 1988, pp. 58–89, p. 79; italics in original.
12 Ibid., p. 82.
13 Ibid.
14 This was a rare example of Lee and his family being involved in anything amiss. See 'Singapore, no discount please, Lee Kuan Yew and his son act to allay misimpressions', *Asiaweek*, 10 May 2001.
15 P. Corrigan and D. Sayer, *The Great Arch: English State Formation as Cultural Revolution*, Oxford: Blackwell, 1985.
16 Gilbert and Nugent, *Everyday Forms of State Formation*, p. 20.
17 K. Marx, *The Economic and Philosophic Manuscript of 1844*, ed. D. J. Struik, New York: International Publishers, 1972, pp. 140–141.
18 F. Jameson, 'Pleasure: A Political Issue', in Formations Collective (ed.), *Formations of Pleasure*, London: Routledge & Kegan Paul, 1983, p. 3.
19 Ibid., p. 4.
20 R. Barthes, *The Pleasure of the Text*, translated by R. Miller, New York: Hill and Wang, 1975.
21 Jameson, 'Pleasure: A Political Issue', p. 8.
22 R. Barthes, *Writing Degree Zero*, New York: Hill and Wang, 1968, pp. 76–77.
23 Ibid., p. 78.
24 Jameson, 'Pleasure: A Political Issue', p. 8.
25 Ibid., p. 9.
26 F. Jameson, *The Political Unconscious*, Ithaca, NY: Cornell University Press, 1981, p. 288.

27 *Star*, 14 April 2002.
28 Blanchot, 'Everyday Speech', p. 13.
29 Ibid.; emphasis added.
30 H. Lefebvre, 'The Everyday and Everydayness', *Yale French Studies* 73, 1987, pp. 7–8.
31 J. Derrida, *Glas*, translated by J. P. Leavey Jr and R. Rand, Lincoln, Nebr.: Nebraska University Press, 1986, p. 161.
32 W. Benjamin, 'Naples', in *One-Way Street and Other Writings*, London: New Left Books, 1979c, p. 176.
33 M. M. Bakhtin, *Rabelais and his World*, Cambridge, Mass.: MIT Press, 1968.
34 T. Eagleton, *Walter Benjamin, or, Towards a Revolutionary Criticism*, London: Verso and New Left Books, 1981.

7 *I Not Stupid*: localism, bad translation, catharsis

1 W. Benjamin, 'The Task of the Translator', in H. Arendt (ed.), *Illuminations*, New York: Schocken Books, 1969a, p. 72.
2 *Business Times*, 19 February 2003.
3 *Time Magazine*, 8 April 2002.
4 *Sunday Star*, 24 February 2002.
5 www.southseattlestar.com/issues/May.14.2003/film.html; accessed 26 March 2004.
6 *Star*, 24 February 2002.
7 Ibid.
8 www.mediacorpsingapore.com/article/feature/view/336/1; accessed 11 July 2004.
9 www.contactsingapore.org.sg/nm/global talent/speeches; accessed 11 July 2004.
10 Benjamin, 'The Task of the Translator', p. 73.
11 Ibid., p. 78.
12 Ibid., p. 76.
13 S. Freud, *Jokes and their Relation to the Unconscious*, Harmondsworth: Penguin, 1976b, p. 137.
14 Ibid., p. 132.
15 www.mediacorpsingapore.com/article/feature/view/336/1; accessed 11 July 2004.

8 The nation after history

1 This and all quotes earlier are from *Straits Times*, 17 September 2001.
2 For a summary of the report and details of Singapore's anti-terrorist operations, see *Asia Times*, 22 November 2003.
3 Ibid. See Chapter 9 for an update of the growth and employment figures.
4 www.oecd.org; accessed 20 April 2004.
5 *Asia Times*, 25 September 2003.
6 *Straits Times*, 14 September 2003.
7 *Asia Times*, 25 September 2003.
8 *Straits Times*, 14 September 2003.

9 Ibid.
10 Ibid.
11 See, for example, F. Fukuyama, *The End of History and the Last Man*, New York; Toronto: Free Press, 1992.
12 *Straits Times*, 14 September 2003.
13 Ibid.
14 M. S. Roth, *Knowing and History: Appropriations of Hegel in Twentieth-Century France*, Ithaca, NY: Cornell University Press, 1988, pp. 103–104.
15 A. Kojève, *Introduction to the Reading of Hegel*, Ithaca, NY: Cornell University Press, 1980, p. 162; italics in original.
16 Ibid.; italics in original.
17 Ibid., p. 159.
18 Ibid., p. 160; italics in original.
19 Ibid., p. 159.
20 Government of Singapore, *White Paper on Constitutional Amendments to Safeguard Financial Assets and the Integrity of the Public Service*, Singapore: Government Press, 1988.
21 K. Tan and P. E. Lam (eds), *Managing Political Change in Singapore: The Elected Presidency*, London; New York: Routledge, 1997, p. xi.
22 *Straits Times*, 15 August 1988.
23 P. E. Lam, 'The Elected Presidency: Towards the Twenty-first Century', in K. Tan and P. E. Lam (eds), *Managing Political Change in Singapore: The Elected Presidency*, London; New York: Routledge, 1997, p. 208.
24 *Straits Times*, 15 August 1988.

9 Epilogue: useless pragmatism

1 Han Suyin, 'Social Change in Asia', *Suloh Nantah: Journal of the English Society*, Nanyang University Singapore, No. 15–16, 1960, p. 2.
2 A. L. Stoler, *Race and the Education of Desire: Foucault's History of Sexuality and Colonial Order of Things*, Durham, NC: Duke University Press, 1995.
3 C. Lefort, *The Political Forms of Modern Society: Bureaucracy, Democracy, Totalitarianism*, ed. J. B. Thompson, Cambridge: Polity Press, 1986.
4 *Star*, 11 April 2004.
5 Ibid.
6 The Singapore Democratic Alliance is a coalition of five opposition parties. The holder of the parliamentary seat, Chiam See Tong, was previously leader of the Singapore Democratic Party on which ticket he first won an election before he was ousted in a power struggle. Chiam is now head of the Singapore People's Party.
7 Reuters, 27 March 2006, www.singapore-window.org/sw06/060327re.htm; accessed 5 April 2006.
8 *Star*, 2 May 2005, www.singapore-window.org/sw06/060402st.htm; accessed 5 April 2006.
9 This has to be seen together with the improvement of the opposition, which contested 47 of the 84 seats, and the fall of PAP votes to 66.6 per cent from 75 per cent in 2001. See *Sydney Morning Herald*, 8 May 2006.
10 Ibid.
11 Ibid.

12 The salaries are calculated according to a complex benchmark system based on the income of the top eight earners of six well-paid professions as reported to the tax authorities. *Star*, 11 May 2003; www.singapore-window.org/sw06/060402st.htm; accessed 5 April 2006.

13 Ibid.

14 Ibid.

15 M. H. Abrams, *A Glossary of Literary Terms*, New York: Holt, Rinehart and Winston, 1957, p. 21.

16 Reuters, 1 February 2006, www.singapore-window.org/sw06/060327re. htm; accessed 5 June 2006.

17 Associated Press, 10 March 2006, www.singapore-window.org/sw06/ 043368ap.htm; accessed 5 June 2006.

References

Abrams, M. H., *A Glossary of Literary Terms*, New York: Holt, Rinehart and Winston, 1957.

Abrams, P., 'Notes on the Difficulty of Studying the State', *Journal of Historical Sociology* 1, 1988, pp. 58–89.

—— 'From Miracle to Crash', *London Review of Books* 20 (8), 1998, pp. 16–18.

Bahrampour, F., 'The Caning of Michael Fay: Can Singapore's Punishment Withstand the Scrutiny of International Law?', *American University Journal of International Law and Policy* 10 (3), 1995, pp. 1075–1108.

Bakhtin, M. M., *Rabelais and his World*, Cambridge, Mass.: MIT Press, 1968.

Baratham, G., *The Caning of Michael Fay*, Singapore: KRP Publishers, 1994.

Barr, M., *Lee Kuan Yew: The Beliefs behind the Man*, Richmond: Curzon, 2000a.

—— 'Lee Kuan Yew and the "Asian Values" Debate', *Asian Studies Review* 24, 2000b, pp. 309–334.

Barthes, R., *Writing Degree Zero*, New York: Hill and Wang, 1968.

—— *The Pleasure of the Text*, translated by R. Miller, New York: Hill and Wang, 1975.

Bauman, Z., 'Modernity and Ambivalence', in M. Featherstone (ed.), *Global Culture: Nationalism, Globalization and Modernity*, London: Sage, 1990.

Benjamin, W., 'The Task of the Translator', in H. Arendt (ed.), *Illuminations*, New York: Schocken Books, 1969a.

—— 'The Storyteller: Reflection on the Works of Nikolai Leskov', in H. Arendt (ed.), *Illuminations*, New York: Schocken Books, 1969b.

—— 'Critique of Violence', in *One-way Street and Other Writings*, London: New Left Books, 1979a.

—— 'Breakfast Room', in *One-way Street and Other Writings*, London: New Left Books, 1979b.

—— 'Naples', in *One-Way Street and Other Writings*, London: New Left Books, 1979c.

Blanchot, M., 'Everyday Speech', *Yale French Studies* 73, 1987, pp. 12–20.

Caruth, C., *Trauma: Explorations in Memory*, Baltimore, Md.: Johns Hopkins University Press, 1995.

Chew, M., *Leaders of Singapore*, Singapore: Resource Press, 1996.

Chua B. H., *Communitarian Ideology and Democracy in Singapore*, London; New York: Routledge, 1995.

—— '"Asian-Values" Discourse and the Resurrection of the Social', *Positions: East Asia Cultures and Critique* 7 (1), 1999, pp. 573–592.

Corrigan, P. and Sayer, D., *The Great Arch: English State Formation as Cultural Revolution*, Oxford: Blackwell, 1985.

Cover, R. M., 'Violence and the Word', *Yale Law Journal* 95, 1986, pp. 1601–1629.

Derrida, J., *Glas*, translated by J. P. Leavey Jr and R. Rand, Lincoln, Nebr.: Nebraska University Press, 1986.

Dostoyevsky, F., *Diary of a Writer*, Vol. 1, New York: Charles Scribner's Sons, 1949.

Drysdale, J., *Singapore: Struggle for Success*, North Sydney: George Allen & Unwin, 1984.

Durkheim, E., 'The Science of Morality', in A. Giddens (ed.), *Emile Durkheim: Selected Writings*, Cambridge: Cambridge University Press, 1972.

Eagleton, T., *Walter Benjamin, or, Towards a Revolutionary Criticism*, London: Verso and New Left Books, 1981.

Enright, D. J., *Conspirators and Poets*, London: Chatto & Windus, 1966.

—— *Memoirs of a Mendicant Professor*, London: Chatto & Windus, 1969.

Foucault, M., *The History of Sexuality: An Introduction*, Harmondsworth: Peregrine, 1984.

Freud, S., *The Interpretation of Dreams*, Harmondsworth: Penguin, 1976a.

—— *Jokes and their Relation to the Unconscious*, Harmondsworth: Penguin, 1976b.

—— *The Freud Reader*, ed. P. Gay, London: Vintage, 1995.

Fukuyama, F., *The End of History and the Last Man*, New York; Toronto: Free Press, 1992.

Gilbert, J. and Nugent, D., *Everyday Forms of State Formation*, Durham, NC: Duke University Press, 1995.

Goh, K. S., *The Practice of Economic Growth*, Singapore: Federal Publications, 1995.

Han, F. K., Tan, S. and Fernandez, W., *Lee Kuan Yew: The Man and his Ideas*, Singapore: Singapore Press Holdings and Times Editions, 1998.

Han Suyin, 'Social Change in Asia', *Suloh Nantah: Journal of the English Society*, Nanyang University Singapore, No. 15–16, 1960, pp. 2–9.

Harper, T. N., *The End of Empire and the Making of Malaya*, Cambridge: Cambridge University Press, 1998.

—— 'Lim Chin Siong and the "Singapore Story"', in J. Q. Tan and K. S. Jomo (eds), *Comet in Our Sky: Lim Chin Siong in History*, Kuala Lumpur: INSAN, 2001.

Held, D., Anderson, J., Gieben, B., Hall, S., Harris, L., Lewis, P., Parker, N. and Turok, B. (eds), *States and Societies*, Oxford: Robertson in association with the Open University, 1983.

Hell, J., *Post-Fascist Fantasies: Psychoanalysis, History, and the Literature of East Germany*, Durham, NC: Duke University Press, 1997.

Heng, G. and Janadas, D., 'State Fatherhood: The Politics of Nationalism, Sexuality, and Race in Singapore', in M. G. Peletz and A. Ong (eds), *Bewitching Women, Pious Men: Gender and Body Politics in Southeast Asia*, Berkeley, Calif.: University of California Press, 1995.

Hingley, R., *The Russian Mind*, London: Bodley Head, 1978.

Hobbes, T., 'Leviathan', in D. Held, J. Anderson, B. Gieben, S. Hall, L. Harris, P. Lewis, N. Parker and B. Turok (eds), *States and Societies*, Oxford: Robertson in association with the Open University, 1983.

Hunt, L., *The Family Romance of the French Revolution*, Berkeley, Calif.: University of California Press, 1992.

Jameson, F., *The Political Unconscious*, Ithaca, NY: Cornell University Press, 1981.

—— 'Pleasure: A Political Issue', in Formations Collective (ed.), *Formations of Pleasure*, London: Routledge & Kegan Paul, 1983.

Joseph, G. M. and Nugent, D., *Everyday Forms of State Formation*, Durham, NC; London: Duke University Press, 1994.

Josey, A., *Lee Kuan Yew: The Struggle for Singapore*, Sydney: Angus & Robertson, 1974.

Kleinman, A., Veena, D. and Lock, M., *Social Suffering*, Berkeley, Calif.: University of California Press, 1997.

Koh, K. L., 'Unnatural Offences: A Widening Scope?', *Review of Judicial and Legal Reform in Singapore between 1990 and 1995*, Singapore: Butterworth Asia, 1996, pp. 362–364.

Kojève, A., *Introduction to the Reading of Hegel*, Ithaca, NY: Cornell University Press, 1980.

Krause, L. B., 'Government as Entrepreneur', in K. S. Sandhu and P. Wheatley (eds), *Management of Success: The Moulding of Modern Singapore*, Singapore: Institute of Southeast Asian Studies, 1989.

Lacan, J., *Ecrites*, translated by A. Sheridan, New York: W. W. Norton, 1977.

Lam, P. E., 'The Elected Presidency: Towards the Twenty-first Century', in K. Tan and P. E. Lam (eds), *Managing Political Change in Singapore: The Elected Presidency*, London; New York: Routledge, 1997.

Lee, K. Y., *Malaysia: Age of Revolution*, Singapore: Ministry of Culture, 1965.

—— *The Singapore Story: Memoirs of Lee Kuan Yew*, Singapore: Prentice Hall, 1998.

—— *From Third World to First: The Singapore Story: 1965–2000*, Singapore: Singapore Press Holdings and Times Editions, 2000.

Lefebvre, H., 'The Everyday and Everydayness', *Yale French Studies* 73, 1987, pp. 7–11.

Lefort, C., *The Political Forms of Modern Society: Bureaucracy, Democracy, Totalitarianism*, ed. J. B. Thompson, Cambridge: Polity Press, 1986.

Lim, M. F. C., 'An Appeal to the Use of the Rod Sparingly: A Dispassionate Analysis of the Use of Caning in Singapore', *Singapore Law Review* 15, 1994, pp. 20–96.

Locke, J., *Two Treatises of Government*, ed. P. Laslette, Cambridge: Cambridge University Press, 1963.

Marx, K., *The Economic and Philosophic Manuscript of 1844*, ed. D. J. Struik, New York: International Publishers, 1972.

Minchin, J., *No Man Is an Island: A Study of Singapore's Lee Kuan Yew*, North Sydney: Allen & Unwin, 1986.

Moore, S. F., *Law as Process: An Anthropological Approach*, London; Boston, Mass.: Routledge & Kegan Paul, 1978.

Nair, C. V. D., *Socialism that Works: The Singapore Way*, Singapore: Federal Publications, 1976.

Pile, S., 'Freud, Dreams and Imaginative Geographies', in A. Elliott (ed.), *Freud 2000*, Carlton: Melbourne University Press, 1998.

Roth, M. S., *Knowing and History: Appropriations of Hegel in Twentieth-Century France*, Ithaca, NY: Cornell University Press, 1988.

Sai, S. M. and Huang, J., 'The "Chinese-Educated" Political Vanguards: Ong Pang Boon, Lee Khoon Choy and Jek Yeun Thong', in P. E. Lam and K. Tan (eds), *Lee's Lieutenants: Singapore's Old Guard*, St Leonards, NSW: Allen & Unwin, 1999.

Said, E. W., *Orientalism*, Harmondsworth: Peregrine, 1978.

Scarry, E., *The Body in Pain: The Making and Unmaking of the World*, New York: Oxford University Press, 1985.

Seow, F. T., *To Catch a Tartar: A Dissident in Lee Kuan Yew's Prison*, New Haven, Conn.: Yale University Southeast Asia Studies, 1994.

—— *The Media Enthralled: Singapore Revisited*, Boulder, Colo.: Lynne Rienner Publishers, 1998.

—— 'The Stranger', in *On Individuality and Social Form*, translated by K. H. Wolff, Chicago, Ill.: Chicago University Press, 1971.

Skinner, Q., *The Foundation of Modern Political Thought*, Vol. 2, Cambridge: Cambridge University Press, 1978.

Smyth, P., *The Economic State and the Welfare State: Australia and Singapore 1955–1975*, Singapore: Centre for Advanced Studies, National University of Singapore, 2000.

Stoler, A. L., *Race and the Education of Desire: Foucault's History of Sexuality and Colonial Order of Things*, Durham, NC: Duke University Press, 1995.

Talkingcock.com, *The Coxford Singlish Dictionary*, Singapore: Angsana Books, 2002.

Tan, K. and Lam, P. E. (eds), *Managing Political Change in Singapore: The Elected Presidency*, London; New York: Routledge, 1997.

Taussig, M., '*Maleficium*: State Fetishism', in E. S. Apter and W. Pietz (eds), *Fetishism as Cultural Discourse*, Ithaca, NY: Cornell University Press, 1993.

Thorpe, M., '"Penetration by Invitation" – A Lecturer at Nanyang – A Memoir', *Economic and Political Weekly*, 1–8 March 1997, pp. 485–491.

Trilling, L., 'Beyond Culture: Essays on Literature and Learning', in L. Trilling (ed.), *The Works of Lionel Trilling*, New York: Harcourt Brace Jovanovich, 1978.

Vincent, A., *Theories of the State*, Oxford: Blackwell, 1987.

Wee, C. J. W.-L., 'The Vanquished: Lim Chin Siong and a Progressivist National Narrative', in P. E. Lam and K. Tan (eds), *Lee's Lieutenants: Singapore's Old Guard*, St Leonards, NSW: Allen & Unwin, 1999.

Yao, S., *Confucian Capitalism: Discourse, Practice and the Myth of Chinese Enterprise*, London: RoutledgeCurzon, 2002.

Yeoh, J., *To Tame a Tiger: The Singapore Story*, Singapore: Wiz-Biz, 1995.

Government reports

Government of Singapore, *White Paper on Constitutional Amendments to Safeguard Financial Assets and the Integrity of the Public Service*, Singapore: Government Press, 1988.

—— *Shared Values*, cmd1, 1991.

—— *Annual Report*, Social Development Unit, 2002.

Index